INTRODUCTION TO STRUCTURAL DYNAMICS AND AEROELASTICITY, SECOND EDITION

This text provides an introduction to structural dynamics and aeroelasticity, with an emphasis on conventional aircraft. The primary areas considered are structural dynamics, static aeroelasticity, and dynamic aeroelasticity. The structural dynamics material emphasizes vibration, the modal representation, and dynamic response. Aeroelastic phenomena discussed include divergence, aileron reversal, airload redistribution, unsteady aerodynamics, flutter, and elastic tailoring. More than one hundred illustrations and tables help clarify the text, and more than fifty problems enhance student learning. This text meets the need for an up-to-date treatment of structural dynamics and aeroelasticity for advanced undergraduate or beginning graduate aerospace engineering students.

Praise from the First Edition

"Wonderfully written and full of vital information by two unequalled experts on the subject, this text meets the need for an up-to-date treatment of structural dynamics and aeroelasticity for advanced undergraduate or beginning graduate aerospace engineering students."

– Current Engineering Practice

"Hodges and Pierce have written this significant publication to fill an important gap in aeronautical engineering education. Highly recommended."

– Choice

"...a welcome addition to the textbooks available to those with interest in aeroelasticity.... As a textbook, it serves as an excellent resource for advanced undergraduate and entry-level graduate courses in aeroelasticity.... Furthermore, practicing engineers interested in a background in aeroelasticity will find the text to be a friendly primer."

– AIAA Bulletin

Dewey H. Hodges is a Professor in the School of Aerospace Engineering at the Georgia Institute of Technology. He is the author of more than 170 refereed journal papers and three books, *Nonlinear Composite Beam Theory* (2006), *Fundamentals of Structural Stability* (2005, with G. J. Simitses), and *Introduction to Structural Dynamics and Aeroelasticity, First Edition* (2002, with G. Alvin Pierce). His research spans the fields of aeroelasticity, dynamics, computational structural mechanics and structural dynamics, perturbation methods, computational optimal control, and numerical analysis.

The late G. Alvin Pierce was Professor Emeritus in the School of Aerospace Engineering at the Georgia Institute of Technology. He is the coauthor of *Introduction to Structural Dynamics and Aeroelasticity, First Edition* with Dewey H. Hodges (2002).

Cambridge Aerospace Series

Editors: Wei Shyy and Vigor Yang

Introduction to Structural Dynamics and Aeroelasticity

Second Edition

Dewey H. Hodges
Georgia Institute of Technology

G. Alvin Pierce
Georgia Institute of Technology

CAMBRIDGE
UNIVERSITY PRESS

32 Avenue of the Americas, New York NY 10013-2473, USA

Cambridge University Press is part of the University of Cambridge.

It furthers the University's mission by disseminating knowledge in the pursuit of education, learning and research at the highest international levels of excellence.

www.cambridge.org
Information on this title: www.cambridge.org/9781107617094

First edition © Dewey H. Hodges and G. Alvin Pierce 2002
Second edition © Dewey H. Hodges and G. Alvin Pierce 2011

First published 2002
Second Edition published 2011
Reprinted 2012

A catalogue record for this publication is available from the British Library

Library of Congress Cataloguing in Publication data
Hodges, Dewey H.
Introduction to structural dynamics and aeroelasticity / Dewey H. Hodges, G. Alvin Pierce. – 2nd ed.
    p.   cm. – (Cambridge aerospace series ; 15)
Includes bibliographical references and index
ISBN 978-1-107-61709-4 Paperback
1. Space vehicles – Dynamics.   2. Aeroelasticity.   I. Pierce, G. Alvin.   II. Title.
TL671.6.H565   2011
629.134'3131–dc22   2011001984

ISBN 978-1-107-61709-4 Paperback

Contents

Figures

Tables

Foreword

From First Edition

A senior-level undergraduate course entitled "Vibration and Flutter" was taught for many years at Georgia Tech under the quarter system. This course dealt with elementary topics involving the static and/or dynamic behavior of structural elements, both without and with the influence of a flowing fluid. The course did not discuss the static behavior of structures in the absence of fluid flow because this is typically considered in courses in structural mechanics. Thus, the course essentially dealt with the fields of structural dynamics (when fluid flow is not considered) and aeroelasticity (when it is).

As the name suggests, structural dynamics is concerned with the vibration and dynamic response of structural elements. It can be regarded as a subset of aeroelasticity, the field of study concerned with interaction between the deformation of an elastic structure in an airstream and the resulting aerodynamic force. Aeroelastic phenomena can be observed on a daily basis in nature (e.g., the swaying of trees in the wind and the humming sound that Venetian blinds make in the wind). The most general aeroelastic phenomena include dynamics, but static aeroelastic phenomena are also important. The course was expanded to cover a full semester, and the course title was appropriately changed to "Introduction to Structural Dynamics and Aeroelasticity."

Aeroelastic and structural-dynamic phenomena can result in dangerous static and dynamic deformations and instabilities and, thus, have important practical consequences in many areas of technology. Especially when one is concerned with the design of modern aircraft and space vehicles—both of which are characterized by the demand for extremely lightweight structures—the solution of many structural dynamics and aeroelasticity problems is a basic requirement for achieving an operationally reliable and structurally optimal system. Aeroelastic phenomena can also play an important role in turbomachinery, civil-engineering structures, wind-energy converters, and even in the sound generation of musical instruments.

Aeroelastic problems may be classified roughly in the categories of response and stability. Although stability problems are the principal focus of the material presented herein, it is not because response problems are any less important. Rather, because the amplitude of deformation is indeterminate in linear stability problems, one may consider an exclusively linear treatment and still manage to solve many practical problems. However, because the amplitude is important in response problems, one is far more likely to need to be concerned with nonlinear behavior when attempting to solve them. Although nonlinear equations come closer to representing reality, the analytical solution of nonlinear equations is problematic, especially in the context of undergraduate studies.

The purpose of this text is to provide an introduction to the fields of structural dynamics and aeroelasticity. The length and scope of the text are intended to be appropriate for a semester-length, senior-level, undergraduate course or a first-year graduate course in which the emphasis is on conventional aircraft. For curricula that provide a separate course in structural dynamics, an ample amount of material has been added to the aeroelasticity chapters so that a full course on aeroelasticity alone could be developed from this text.

This text was built on the foundation provided by Professor Pierce's course notes, which had been used for the "Vibration and Flutter" course since the 1970s. After Professor Pierce's retirement in 1995, when the responsibility for the course was transferred to Professor Hodges, the idea was conceived of turning the notes into a more substantial text. This process began with the laborious conversion of Professor Pierce's original set of course notes to LaTeX format in the fall of 1997. The authors are grateful to Margaret Ojala, who was at that time Professor Hodges's administrative assistant and who facilitated the conversion. Professor Hodges then began the process of expanding the material and adding problems to all chapters. Some of the most substantial additions were in the aeroelasticity chapters, partly motivated by Georgia Tech's conversion to the semester system. Dr. Mayuresh J. Patil,[1] while he was a Postdoctoral Fellow in the School of Aerospace Engineering, worked with Professor Hodges to add material on aeroelastic tailoring and unsteady aerodynamics mainly during the academic year 1999–2000. The authors thank Professor David A. Peters of Washington University for his comments on the section that treats unsteady aerodynamics. Finally, Professor Pierce, while enjoying his retirement and building a new house and amid a computer-hardware failure and visits from grandchildren, still managed to add material on the history of aeroelasticity and on the k and p-k methods in the early summer of 2001.

Dewey H. Hodges and G. Alvin Pierce
Atlanta, Georgia
June 2002

---

[1] Presently, Dr. Patil is Associate Professor in the Department of Aerospace and Ocean Engineering at Virginia Polytechnic and State University.

Addendum for Second Edition

Plans for the second edition were inaugurated in 2007, when Professor Pierce was still alive. All his colleagues at Georgia Tech and in the technical community at large were saddened to learn of his death in November 2008. Afterward, plans for the second edition were somewhat slow to develop.

The changes made for the second edition include additional material along with extensive reorganization. Instructors may choose to omit certain sections without breaking the continuity of the overall treatment. Foundational material in mechanics and structures was somewhat expanded to make the treatment more self-contained and collected into a single chapter. It is hoped that this new organization will facilitate students who do not need this review to easily skip it, and that students who do need it will find it convenient to have it consolidated into one relatively short chapter. A discussion of stability is incorporated, along with a review of how single-degree-of-freedom systems behave as key parameters are varied. More detail is added for obtaining numerical solutions of characteristic equations in structural dynamics. Students are introduced to finite-element structural models, making the material more commensurate with industry practice. Material on control reversal in static aeroelasticity has been added. Discussion on numerical solution of the flutter determinant via Mathematica™ replaces the method presented in the first edition for interpolating from a set of candidate reduced frequencies. The treatment of flutter analysis based on complex eigenvalues is expanded to include an unsteady-aerodynamics model that has its own state variables. Finally, the role of flight-testing and certification is discussed. It is hoped that the second edition will not only maintain the text's uniqueness as an undergraduate-level treatment of the subject, but that it also will prove to be more useful in a first-year graduate course.

Dewey H. Hodges
Atlanta, Georgia

INTRODUCTION TO STRUCTURAL DYNAMICS AND AEROELASTICITY, SECOND EDITION

1   Introduction

"Aeroelasticity" is the term used to denote the field of study concerned with the interaction between the deformation of an elastic structure in an airstream and the resulting aerodynamic force. The interdisciplinary nature of the field is best illustrated by Fig. 1.1, which originated with Professor A. R. Collar in the 1940s. This triangle depicts interactions among the three disciplines of aerodynamics, dynamics, and elasticity. Classical aerodynamic theories provide a prediction of the forces acting on a body of a given shape. Elasticity provides a prediction of the shape of an elastic body under a given load. Dynamics introduces the effects of inertial forces. With the knowledge of elementary aerodynamics, dynamics, and elasticity, students are in a position to look at problems in which two or more of these phenomena interact. The field of flight mechanics involves the interaction between aerodynamics and dynamics, which most undergraduate students in an aeronautics/aeronautical engineering curriculum have studied in a separate course by their senior year. This text considers the three remaining areas of interaction, as follows:

- between elasticity and dynamics (i.e., structural dynamics)
- between aerodynamics and elasticity (i.e., static aeroelasticity)
- among all three (i.e., dynamic aeroelasticity)

Because of their importance to aerospace system design, these areas are also appropriate for study in an undergraduate aeronautics/aeronautical engineering curriculum. In aeroelasticity, one finds that the loads depend on the deformation (i.e., aerodynamics) and that the deformation depends on the loads (i.e., structural mechanics/dynamics); thus, one has a coupled problem. Consequently, prior study of all three constituent disciplines is necessary before a study in aeroelasticity can be undertaken. Moreover, a study in structural dynamics is helpful in developing concepts that are useful in solving aeroelasticity problems, such as the modal representation.

It is of interest that aeroelastic phenomena played a major role throughout the history of powered flight. The Wright brothers utilized controlled warping of the wings on their Wright Flyer in 1903 to achieve lateral control. This was essential to their success in achieving powered flight because the aircraft was laterally unstable due to the significant anhedral of the wings. Earlier in 1903, Samuel Langley made

1

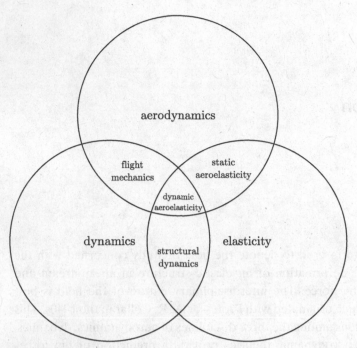

Figure 1.1. Schematic of the field of aeroelasticity

two attempts to achieve powered flight from the top of a houseboat on the Potomac River. His efforts resulted in catastrophic failure of the wings caused by their being overly flexible and overloaded. Such aeroelastic phenomena, including torsional divergence, were major factors in the predominance of the biplane design until the early 1930s, when "stressed-skin" metallic structural configurations were introduced to provide adequate torsional stiffness for monoplanes.

The first recorded and documented case of flutter in an aircraft occurred in 1916. The Handley Page O/400 bomber experienced violent tail oscillations as the result of the lack of a torsion-rod connection between the port and starboard elevators—an absolute design requirement of today. The incident involved a dynamic twisting of the fuselage to as much as 45 degrees in conjunction with an antisymmetric flapping of the elevators. Catastrophic failures due to aircraft flutter became a major design concern during the First World War and remain so today. R. A. Frazer and W. J. Duncan at the National Physical Laboratory in England compiled a classic document on this subject entitled, "The Flutter of Aeroplane Wings" as R&M 1155 in August 1928. This small document (about 200 pages) became known as "The Flutter Bible." Their treatment for the analysis and prevention of the flutter problem laid the groundwork for the techniques in use today.

Another major aircraft-design concern that may be classified as a static-aeroelastic phenomenon was experienced in 1927 by the Bristol Bagshot, a twin-engine, high-aspect-ratio English aircraft. As the speed was increased, the aileron effectiveness decreased to zero and then became negative. This loss and reversal of aileron control is commonly known today as "aileron reversal." The incident was successfully analyzed and design criteria were developed for its prevention by Roxbee Cox and Pugsley at the Royal Aircraft Establishment in the early 1930s. Although aileron reversal generally does not lead to a catastrophic failure, it can be dangerous and therefore is an essential design concern. It is of interest that during this period of the early 1930s, it was Roxbee Cox and Pugsley who proposed the name "aeroelasticity" to describe these phenomena, which are the subject of this text.

In the design of aerospace vehicles, aeroelastic phenomena can result in a full spectrum of behavior from the near benign to the catastrophic. At the near-benign end of the spectrum, one finds passenger and pilot discomfort. One moves from there to steady-state and transient vibrations that slowly cause an aircraft structure to suffer fatigue damage at the microscopic level. At the catastrophic end, aeroelastic instabilities can quickly destroy an aircraft and result in loss of human life without warning. Aeroelastic problems that need to be addressed by aerospace system designers can be mainly static in nature—meaning that inertial forces do not play a significant role—or they can be strongly influenced by inertial forces. Although not the case in general, the analysis of some aeroelastic phenomena can be undertaken by means of small-deformation theories. Aeroelastic phenomena may strongly affect the performance of an aircraft, positively or negatively. They also may determine whether its control surfaces perform their intended functions well, poorly, or even in the exact opposite manner of that which they are intended to do. It is clear then that all of these studies have important practical consequences in many areas of aerospace technology. The design of modern aircraft and space vehicles is characterized by the demand for extremely lightweight structures. Therefore, the solution of many aeroelastic problems is a basic requirement for achieving an operationally reliable and structurally optimal system. Aeroelastic phenomena also play an important role in turbomachinery, in wind-energy converters, and even in the sound generation of musical instruments.

The most commonly posed problems for the aeroelastician are stability problems. Although the elastic moduli of a given structural member are independent of the speed of the aircraft, the aerodynamic forces strongly depend on it. It is therefore not difficult to imagine scenarios in which the aerodynamic forces "overpower" the elastic restoring forces. When this occurs in such a way that inertial forces have little effect, we refer to this as a static-aeroelastic instability—or "divergence." In contrast, when the inertial forces are important, the resulting dynamic instability is called "flutter." Both divergence and flutter can be catastrophic, leading to sudden destruction of a vehicle. Thus, it is vital for aircraft designers to know how to design lifting surfaces that are free of such problems. Most of the treatment of aeroelasticity in this text is concerned with stability problems.

Much of the rest of the field of aeroelasticity involves a study of aircraft response in flight. Static-aeroelastic response problems constitute a special case in which inertial forces do not contribute and in which one may need to predict the lift developed by an aircraft of given configuration at a specified angle of attack or determine the maximum load factor that such an aircraft can sustain. Also, problems of control effectiveness and aileron reversal fall in this category. When inertial forces are important, one may need to know how the aircraft reacts in turbulence or in gusts. Another important phenomenon is buffeting, which is characterized by transient vibration induced by wakes behind wings, nacelles, or other aircraft components.

All of these problems are treatable within the context of a linear analysis. Mathematically, linear problems in aeroelastic response and stability are complementary. That is, instabilities are predictable from examining the situations under which homogeneous equations possess nontrivial solutions. Response problems, however, are generally based on the solution of inhomogeneous equations. When the system becomes unstable, a solution to the inhomogeneous equations ceases to exist, whereas the homogeneous equations and boundary conditions associated with a stable conguration do not have a nontrivial solution.

Unlike the predictions from linear analyses, in actual aircraft, it is possible for self-excited oscillations to develop, even at speeds less than the flutter speed. Moreover, large disturbances can "bump" a system that is predicted to be stable by linear analyses into a state of large oscillatory motion. Both situations can lead to steady-state periodic oscillations for the entire system, called "limit-cycle oscillations." In such situations, there can be fatigue problems leading to concerns about the life of certain components of an aircraft as well as passenger comfort and pilot endurance. To capture such behavior in an analysis, the aircraft must be treated as a nonlinear system. Although of great practical importance, nonlinear analyses are beyond the scope of this textbook.

The organization of the text is as follows. The fundamentals of mechanics are reviewed in Chapter 2. Later chapters frequently refer to this chapter for the formulations embodied therein, including the dynamics of particles and rigid bodies along with analyses of strings and beams as examples of simple structural elements. Finally, the behavior of single-degree-of-freedom systems is reviewed along with a physically motivated discussion of stability.

To describe the dynamic behavior of conventional aircraft, the topic of structural dynamics is introduced in Chapter 3. This is the study of dynamic properties of continuous elastic configurations, which provides a means of analytically representing a flight vehicle's deformed shape at any instant of time. We begin with simple systems, such as vibrating strings, and move up in complexity to beams in torsion and finally to beams in bending. The introduction of the modal representation and its subsequent use in solving aeroelastic problems is the main emphasis of Chapter 3. A brief introduction to the methods of Ritz and Galerkin is also included.

Chapter 4 addresses static aeroelasticity. The chapter is concerned with static instabilities, steady airloads, and control-effectiveness problems. Again, we begin with simple systems, such as elastically restrained rigid wings. We move to wings in torsion and swept wings in bending and torsion and then finish the chapter with a treatment of swept composite wings undergoing elastically coupled bending and torsional deformation.

Finally, Chapter 5 discusses aeroelastic flutter, which is associated with dynamic-aeroelastic instabilities due to the mutual interaction of aerodynamic, elastic, and inertial forces. A generic lifting-surface analysis is first presented, followed by illustrative treatments involving simple "typical-section models." Engineering solution methods for flutter are discussed, followed by a brief presentation of unsteady-aerodynamic theories, both classical and modern. The chapter concludes with an application of the modal representation to the flutter analysis of flexible wings, a discussion of the flutter-boundary characteristics of conventional aircraft, and an overview of how structural dynamics and aeroelasticity impact flight tests and certification. It is important to note that central to our study in the final two chapters are the phenomena of divergence and flutter, which typically result in catastrophic failure of the lifting surface and may lead to subsequent destruction of the flight vehicle.

An appendix is included in which Lagrange's equations are derived and illustrated, as well as references for structural dynamics and aeroelasticity.

Mechanics Fundamentals

> Although to penetrate into the intimate mysteries of nature and thence to learn the
> true causes of phenomena is not allowed to us, nevertheless it can happen that a
> certain fictive hypothesis may suffice for explaining many phenomena.
>
> —Leonard Euler

As discussed in Chapter 1, both structural dynamics and aeroelasticity are built on
the foundations of dynamics and structural mechanics. Therefore, in this chapter, we
review the fundamentals of mechanics for particles, rigid bodies, and simple structures such as strings and beams. The review encompasses laws of motion, expressions
for energy and work, and background assumptions. The chapter concludes with a
brief discussion of the behavior of single-degree-of-freedom systems and the notion
of stability.

The field of structural dynamics addresses the dynamic deformation behavior
of continuous structural configurations. In general, load-deflection relationships are
nonlinear, and the deflections are not necessarily small. In this chapter, to facilitate
tractable, analytical solutions, we restrict our attention to linearly elastic systems
undergoing small deflections—conditions that typify most flight-vehicle operations.

However, some level of geometrically nonlinear theory is necessary to arrive at
a set of linear equations for strings, membranes, helicopter blades, turbine blades,
and flexible rods in rotating spacecraft. Among these problems, only strings are
discussed herein. Indeed, linear equations of motion for free vibration of strings
cannot be obtained without initial consideration and subsequent careful elimination
of nonlinearities.

The treatment goes beyond material generally presented in textbooks when
it delves into the modeling of composite beams. By virtue of the inclusion of this
section, readers obtain more than a glimpse of the physical phenomena associated
with these evermore pervasive structural elements to the point that such beams can
be treated in a simple fashion suitable for use in aeroelastic tailoring (see Chapter 4).
The treatment follows along with the spirit of Euler's quotation: in mechanics, we
seek to make certain assumptions (i.e., fictive hypotheses) that although they do
not necessarily provide knowledge of true causes, they do afford us a mathematical

model that is useful for analysis and design. The usefulness of such models is only as good as can be validated against experiments or models of higher fidelity. For example, defining a beam as a slender structural element in which one dimension is much larger than the other two, we observe that many aircraft wings do not have the geometry of a beam. If the aspect ratio is sufficiently large, however, a beam model may suffice to describe the overall behavioral characteristics of a wing.

2.1 Particles and Rigid Bodies

The simplest dynamical systems are particles. The particle is idealized as a "point-mass," meaning that it takes up no space even though it has nonzero mass. The position vector of a particle in a Cartesian frame can be characterized in terms of its three Cartesian coordinates—for example, x, y, and z. Particles have velocity and acceleration but they do not have angular velocity or angular acceleration. Introducing three unit vectors, $\hat{\mathbf{i}}, \hat{\mathbf{j}}$, and $\hat{\mathbf{k}}$, which are regarded as fixed in a Cartesian frame F, one may write the position vector of a particle Q relative to a point O fixed in F as

$$\mathbf{p}_Q = x\hat{\mathbf{i}} + y\hat{\mathbf{j}} + z\hat{\mathbf{k}} \tag{2.1}$$

The velocity of Q in F can then be written as a time derivative of the position vector in which one regards the unit vectors as fixed (i.e., having zero time derivatives) in F, so that

$$\mathbf{v}_Q = \dot{x}\hat{\mathbf{i}} + \dot{y}\hat{\mathbf{j}} + \dot{z}\hat{\mathbf{k}} \tag{2.2}$$

Finally, the acceleration of Q in F is given by

$$\mathbf{a}_Q = \ddot{x}\hat{\mathbf{i}} + \ddot{y}\hat{\mathbf{j}} + \ddot{z}\hat{\mathbf{k}} \tag{2.3}$$

2.1.1 Newton's Laws

An inertial frame is a frame of reference in which Newton's laws are valid. The only way to ascertain whether a particular frame is sufficiently close to being inertial is by comparing calculated results with experimental data. These laws may be stated as follows:

1st Particles with zero resultant force acting on them move with constant velocity in an inertial frame.

2nd The resultant force on a particle is equal to its mass times its acceleration in an inertial frame. In other words, this acceleration is defined as in Eq. (2.3), with the frame F being an inertial frame.

3rd When a particle P exerts a force on another particle Q, Q simultaneously exerts a force on P with the same magnitude and line of action but in the opposite direction. This law is often simplified as the sentence: "To every action, there is an equal and opposite reaction."

2.1.2 Euler's Laws and Rigid Bodies

Euler generalized Newton's laws to systems of particles, including rigid bodies. A rigid body B may be regarded kinematically as a reference frame. It is easy to show that the position of every point in B is determined in a frame of reference F if (a) the position of any point fixed in B, such as its mass center C, is known in the frame of reference F; and (b) the orientation of B is known in F.

Euler's first law for a rigid body simply states that the resultant force acting on a rigid body is equal to its mass times the acceleration of the body's mass center in an inertial frame. Euler's second law is more involved and may be stated in several ways. The two ways used most commonly in this text are as follows:

- The sum of torques about the mass center C of a rigid body is equal to the time rate of change in F of the body's angular momentum in F about C, with F being an inertial frame.
- The sum of torques about a point O that is fixed in the body and is also inertially fixed is equal to the time rate of change in F of the body's angular momentum in F about O, with F being an inertial frame. We subsequently refer to O as a "pivot."

Consider a rigid body undergoing two-dimensional motion such that the mass center C moves in the x-y plane and the body has rotational motion about the z axis. Assuming the body to be "balanced" in that the products of inertia $I_{xz} = I_{yz} = 0$, Euler's second law can be simplified to the scalar equation

$$T_C = I_C \ddot{\theta} \tag{2.4}$$

where T_C is the moment of all forces about the z axis passing through C, I_C is the moment of inertia about C, and $\ddot{\theta}$ is the angular acceleration in an inertial frame of the body about z. This equation also holds if C is replaced by O.

2.1.3 Kinetic Energy

The kinetic energy K of a particle Q in F can be written as

$$K = \frac{m}{2} \mathbf{v}_Q \cdot \mathbf{v}_Q \tag{2.5}$$

where m is the mass of the particle and $\mathbf{v}_Q$ is the velocity of Q in F. To use this expression for the kinetic energy in mechanics, F must be an inertial frame.

The kinetic energy of a rigid body B in F can be written as

$$K = \frac{m}{2} \mathbf{v}_C \cdot \mathbf{v}_C + \frac{1}{2} \boldsymbol{\omega}_B \cdot \underline{\mathbf{I}}_C \cdot \boldsymbol{\omega}_B \tag{2.6}$$

where m is the mass of the body, $\underline{\mathbf{I}}_C$ is the inertia tensor of B about C, $\mathbf{v}_C$ is the velocity of C in F, and $\boldsymbol{\omega}_B$ is the angular velocity of B in F. In two-dimensional motion of a balanced body, we may simplify this to

$$K = \frac{m}{2} \mathbf{v}_C \cdot \mathbf{v}_C + \frac{I_C}{2} \dot{\theta}^2 \tag{2.7}$$

where I_C is the moment of inertia of B about C about z, $\dot{\theta}$ is the angular velocity of B in F about z, and z is an axis perpendicular to the plane of motion. A similar equation also holds if C is replaced by O, a pivot, such that

$$K = \frac{I_O}{2}\dot{\theta}^2 \tag{2.8}$$

where I_O is the moment of inertia of B about an axis z passing through O. To use these expressions for kinetic energy in mechanics, F must be an inertial frame.

2.1.4 Work

The work W done in a reference frame F by a force $\mathbf{F}$ acting at a point Q, which may be either a particle or a point on a rigid body, may be written as

$$W = \int_{t_1}^{t_2} \mathbf{F} \cdot \mathbf{v}_Q dt \tag{2.9}$$

where $\mathbf{v}_Q$ is the velocity of Q in F, and t_1 and t_2 are arbitrary fixed times. When there are contact and distance forces acting on a rigid body, we may express the work done by all such forces in terms of their resultant $\mathbf{R}$, acting at C, and the total torque $\mathbf{T}$ of all such forces about C, such that

$$W = \int_{t_1}^{t_2} (\mathbf{R} \cdot \mathbf{v}_C + \mathbf{T} \cdot \boldsymbol{\omega}_B)\, dt \tag{2.10}$$

The most common usage of these formulae in this text is the calculation of virtual work (i.e., the work done by applied forces through a virtual displacement).

2.1.5 Lagrange's Equations

There are several occasions to make use of Lagrange's equations when calculating the forced response of structural systems. Lagrange's equations are derived in the Appendix and can be written as

$$\frac{d}{dt}\left(\frac{\partial L}{\partial \dot{\xi}_i}\right) - \frac{\partial L}{\partial \xi_i} = \Xi_i \qquad (i = 1, 2, \ldots) \tag{2.11}$$

where $L = K - P$ is called the "Lagrangean"—that is, the difference between the total kinetic energy, K, and the total potential energy, P, of the system. The generalized coordinates are ξ_i; the term on the right-hand side, Ξ_i, is called the "generalized force." The latter represents the effects of all nonconservative forces, as well as any conservative forces that are not treated in the total potential energy.

Under many circumstances, the kinetic energy can be represented as a function of only the coordinate rates so that

$$K = K(\dot{\xi}_1, \dot{\xi}_2, \dot{\xi}_3, \ldots) \tag{2.12}$$

The potential energy P consists of contributions from strain energy, discrete springs, gravity, applied loads (conservative only), and so on. The potential energy is a function of only the coordinates themselves; that is

$$P = P(\xi_1, \xi_2, \xi_3, \ldots) \tag{2.13}$$

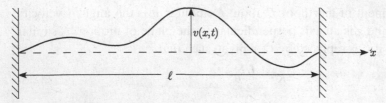

Figure 2.1. Schematic of vibrating string

Thus, Lagrange's equations can be written as

$$\frac{d}{dt}\left(\frac{\partial K}{\partial \dot{\xi}_i}\right) + \frac{\partial P}{\partial \xi_i} = \Xi_i \qquad (i = 1, 2, \ldots) \tag{2.14}$$

2.2 Modeling the Dynamics of Strings

Among the continuous systems to be considered in other chapters, the string is the simplest. Typically, by this time in their undergraduate studies, most students have had some exposure to the solution of string-vibration problems. Here, we present for future reference a derivation of the governing equation, the potential energy, and the kinetic energy along with the virtual work of an applied distributed transverse force.

2.2.1 Equations of Motion

A string of initial length ℓ_0 is stretched in the x direction between two walls separated by a distance $\ell > \ell_0$. The string tension, $T(x, t)$, is considered high, and the transverse displacement $v(x, t)$ and slope $\beta(x, t)$ are eventually regarded as small. At any given instant, this system can be illustrated as in Fig. 2.1. To describe the dynamic behavior of this system, the forces acting on a differential length dx of the string can be illustrated by Fig. 2.2. Note that the longitudinal displacement $u(x, t)$, transverse displacement, slope, and tension at the right end of the differential element are

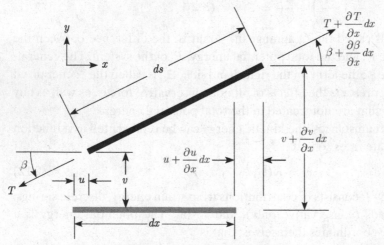

Figure 2.2. Differential element of string showing displacement components and tension force

represented as a Taylor series expansion of the values at the left end. Because the string segment is of a differential length that can be arbitrarily small, the series is truncated by neglecting terms of the order of dx^2 and higher.

Neglecting gravity and any other applied loads, two equations of motion can be formed by resolving the tension forces in the x and y directions and setting the resultant force on the differential element equal to its mass $m\,dx$ times the acceleration of its mass center. Neglecting higher-order differentials, we obtain the equations of motion as

$$\frac{\partial}{\partial x}[T\cos(\beta)] = m\frac{\partial^2 u}{\partial t^2} \qquad \frac{\partial}{\partial x}[T\sin(\beta)] = m\frac{\partial^2 v}{\partial t^2} \tag{2.15}$$

where for a string homogeneous over its cross section

$$m = \iint_A \rho\, d\mathcal{A} = \rho A \tag{2.16}$$

is the mass per unit length. From Fig. 2.2, ignoring second and higher powers of dx and letting $ds = (1+\epsilon)dx$ where ϵ is the elongation, we can identify

$$\cos(\beta) = \frac{1}{1+\epsilon}\left(1+\frac{\partial u}{\partial x}\right)$$

$$\sin(\beta) = \frac{1}{1+\epsilon}\frac{\partial v}{\partial x} \tag{2.17}$$

Noting that $\cos^2(\beta) + \sin^2(\beta) = 1$, we can find the elongation ϵ as

$$\epsilon = \frac{\partial s}{\partial x} - 1 = \sqrt{\left(1+\frac{\partial u}{\partial x}\right)^2 + \left(\frac{\partial v}{\partial x}\right)^2} - 1 \tag{2.18}$$

Finally, considering the string as homogeneous, isotropic, and linearly elastic, we can write the tension force as a linear function of the elongation, so that

$$T = EA\epsilon \tag{2.19}$$

where EA is the constant longitudinal stiffness of the string. This completes the system of nonlinear equations that govern the vibration of the string. To develop analytical solutions, we must simplify these equations.

Let us presuppose the existence of a static-equilibrium solution of the string deflection so that

$$u(x,t) = \bar{u}(x)$$
$$v(x,t) = 0$$
$$\beta(x,t) = 0 \tag{2.20}$$
$$\epsilon(x,t) = \bar{\epsilon}(x)$$
$$T(x,t) = \bar{T}(x)$$

We then find that such a solution exists and that if $\overline{u}(0) = 0$

$$\overline{T}(x) = T_0$$

$$\overline{\epsilon}(x) = \epsilon_0 = \frac{T_0}{EA} = \frac{\delta}{\ell_0} \tag{2.21}$$

$$\overline{u}(x) = \epsilon_0 x$$

where T_0 and ϵ_0 are constants and $\delta = \ell - \ell_0$ is the change in the length of the string between its stretched and unstretched states.

If the steady-state tension T_0 is sufficiently high, the perturbation deflections about the static-equilibrium solution are very small. Thus, we can assume

$$u(x, t) = \overline{u}(x) + \hat{u}(x, t)$$

$$v(x, t) = \hat{v}(x, t)$$

$$\beta(x, t) = \hat{\beta}(x, t) \tag{2.22}$$

$$\epsilon(x, t) = \overline{\epsilon}(x) + \hat{\epsilon}(x, t)$$

$$T(x, t) = \overline{T}(x) + \hat{T}(x, t)$$

where the $(\hat{\ })$ quantities are taken to be infinitesimally small. Furthermore, from the second of Eqs. (2.17), we can determine $\hat{\beta}$ in terms of the other quantities; that is

$$\hat{\beta} = \frac{1}{1 + \epsilon_0} \frac{\partial \hat{v}}{\partial x} \tag{2.23}$$

Substituting the perturbation expressions of Eqs. (2.22) and (2.23) into Eqs. (2.15) while ignoring all squares and products of the $(\hat{\ })$ quantities, we find that the equations of motion can be reduced to two linear partial differential equations

$$EA\frac{\partial^2 \hat{u}}{\partial x^2} = m\frac{\partial^2 \hat{u}}{\partial t^2}$$

$$\frac{T_0}{1 + \epsilon_0} \frac{\partial^2 \hat{v}}{\partial x^2} = m\frac{\partial^2 \hat{v}}{\partial t^2} \tag{2.24}$$

Thus, the two nonlinear equations of motion in Eqs. (2.15) for the free vibration of a string have been reduced to two uncoupled linear equations: one for longitudinal vibration and the other for transverse vibrations. Because it is typically true that $EA \gg T_0$, longitudinal motions have much smaller amplitudes and much higher natural frequencies; thus, they are not usually of interest. Moreover, the fact that $EA \gg T_0$ leads to the observations that $\epsilon_0 \ll 1$ and $\delta \ll \ell_0$ (see Eqs. 2.21). Thus, the transverse motion is governed by

$$T_0\frac{\partial^2 \hat{v}}{\partial x^2} = m\frac{\partial^2 \hat{v}}{\partial t^2} \tag{2.25}$$

For convenience, we drop the $(\ )$s and the subscript, thereby yielding the usual equation for string vibration found in texts on vibration

$$T\frac{\partial^2 v}{\partial x^2} = m\frac{\partial^2 v}{\partial t^2} \tag{2.26}$$

This is called the one-dimensional "wave equation," and it governs the structural dynamic behavior of the string in conjunction with boundary conditions and initial conditions. The fact that the equation is of second order both temporally and spatially indicates that two boundary conditions and two initial conditions need to be specified. The boundary conditions at the ends of the string correspond to zero displacement, as described by

$$v(0, t) = v(\ell, t) = 0 \tag{2.27}$$

where it is noted that the distinction between ℓ_0 and ℓ is no longer relevant. The general solution to the wave equation with these homogeneous boundary conditions comprises a simple eigenvalue problem; the solution, along with a treatment of the initial conditions, is in Section 3.1.

2.2.2 Strain Energy

To solve problems involving the forced response of strings using Lagrange's equation, we need an expression for the strain energy, which is caused by extension of the string, viz.

$$P = \frac{1}{2}\int_0^{\ell_0} EA\epsilon^2 dx \tag{2.28}$$

where, as before

$$\epsilon = \frac{\partial s}{\partial x} - 1 = \sqrt{\left(1 + \frac{\partial u}{\partial x}\right)^2 + \left(\frac{\partial v}{\partial x}\right)^2} - 1 \tag{2.29}$$

and the original length is ℓ_0. To pick up all of the linear terms in Lagrange's equations, we must include all terms in the energy up through the second power of the unknowns. Taking the pertinent unknowns to be perturbations relative to the stretched but undeflected string, we can again write

$$\epsilon(x, t) = \bar{\epsilon}(x) + \hat{\epsilon}(x, t)$$
$$u(x, t) = \bar{u}(x) + \hat{u}(x, t) \tag{2.30}$$
$$v(x, t) = \hat{v}(x, t)$$

For EA equal to a constant, the strain energy is

$$P = \frac{EA}{2}\int_0^{\ell_0} \left(\bar{\epsilon}^2 + 2\bar{\epsilon}\hat{\epsilon} + \hat{\epsilon}^2\right) dx \tag{2.31}$$

From Eqs. (2.21), we know that $\bar{T} = T_0$ and $\bar{\epsilon} = \epsilon_0$, where T_0 and ϵ_0 are constants. Thus, the first term of P is a constant and can be ignored. Because $T_0 = EA\epsilon_0$, the strain energy simplifies to

$$P = T_0 \int_0^{\ell_0} \hat{\epsilon} \, dx + \frac{EA}{2} \int_0^{\ell_0} \hat{\epsilon}^2 \, dx \tag{2.32}$$

Using Eqs. (2.29) and (2.30), we find that the longitudinal strain becomes

$$\hat{\epsilon} = \frac{\partial \hat{u}}{\partial x} + \frac{1}{2(1 + \epsilon_0)} \left(\frac{\partial \hat{v}}{\partial x} \right)^2 + \cdots \tag{2.33}$$

where the ellipsis refers to terms of third and higher degree in the spatial partial derivatives of $\hat{u}$ and $\hat{v}$. Then, when we drop all terms that are of third and higher degree in the spatial partial derivatives of $\hat{u}$ and $\hat{v}$, the strain energy becomes

$$P = T_0 \int_0^{\ell_0} \frac{\partial \hat{u}}{\partial x} dx + \frac{T_0}{2(1 + \epsilon_0)} \int_0^{\ell_0} \left(\frac{\partial \hat{v}}{\partial x} \right)^2 dx + \frac{EA}{2} \int_0^{\ell_0} \left(\frac{\partial \hat{u}}{\partial x} \right)^2 dx + \cdots \tag{2.34}$$

Assuming $\hat{u}(0) = \hat{u}(\ell_0) = 0$, we find that the first term vanishes. Because perturbations of the transverse deflections are the unknowns in which we are most interested, and because perturbations of the longitudinal displacements are uncoupled from these and involve oscillations with much higher frequency, we do not need the last term. This leaves only the second term. As before, noting that $\epsilon_0 \ll 1$ and dropping the $\hat{}$ and subscripts for convenience, we obtain the potential energy for a vibrating string

$$P = \frac{T}{2} \int_0^{\ell} \left(\frac{\partial v}{\partial x} \right)^2 dx \tag{2.35}$$

as found in vibration texts.

In any continuous system—whether a string, beam, plate, or shell—we may account for an attached spring by regarding it as an external force and thus determining its contribution to the generalized forces. Such attached springs may be either discrete (i.e., at a point) or distributed. Conversely, we may treat them as added parts of the system by including their potential energies (see Chapter 3, Problem 5). Be careful to not count forces twice; the same is true for any other entity as well.

2.2.3 Kinetic Energy

To solve problems involving the forced response of strings using Lagrange's equation, we also need the kinetic energy. The kinetic energy for a differential length of string is

$$dK = \frac{m}{2} \left[\left(\frac{\partial u}{\partial t} \right)^2 + \left(\frac{\partial v}{\partial t} \right)^2 \right] dx \tag{2.36}$$

Recalling that the longitudinal displacement u was shown previously to be less significant than the transverse displacement v and to uncouple from it for small-perturbation motions about the static-equilibrium state, we may now express the kinetic energy of the whole string over length ℓ as

$$K = \frac{1}{2} \int_0^\ell m \left(\frac{\partial v}{\partial t} \right)^2 dx \qquad (2.37)$$

2.2.4 Virtual Work of Applied, Distributed Force

To solve problems involving the forced response of strings using Lagrange's equation, we also need a general expression for the virtual work of all forces not accounted for in the potential energy. These applied forces and moments are identified most commonly as externally applied loads, which may or may not be a function of the response. They also include any dissipative loads, such as those from dampers. To determine the contribution of distributed transverse loads, denoted by $f(x, t)$, the virtual work may be computed as the work done by applied forces through a virtual displacement, viz.

$$\overline{\delta W} = \int_0^\ell f(x, t) \delta v(x, t) dx \qquad (2.38)$$

where the virtual displacement δv also may be thought of as the Lagrangean variation of the displacement field. Such a variation may be thought of as an increment of the displacement field that satisfies all geometric constraints.

2.3 Elementary Beam Theory

Now that we have considered the fundamental aspects of structural dynamics analysis for strings, these same concepts are applied to the dynamics of beam torsional and bending deformation. The beam has many more of the characteristics of typical aeronautical structures. Indeed, high-aspect-ratio wings and helicopter rotor blades are frequently idealized as beams, especially in conceptual and preliminary design. Even for low-aspect-ratio wings, although a plate model may be more realistic, the bending and torsional deformation can be approximated by use of beam theory with adjusted stiffness coefficients.

2.3.1 Torsion

In an effort to retain a level of simplicity that promotes tractability, the St. Venant theory of torsion is used and the problem is idealized to the extent that torsion is uncoupled from transverse deflections. The torsional rigidity, denoted by $\overline{GJ}$, is taken as given and may vary with x. For homogeneous and isotropic beams, $\overline{GJ} = GJ$, where G denotes the shear modulus and J is a constant that depends only on the geometry of the cross section. To be uncoupled from bending and other types of deformation, the x axis must be along the elastic axis and also must coincide with the locus of cross-sectional mass centroids. For isotropic beams, the elastic axis is along the locus of cross-sectional shear centers.

For such beams, J can be determined by solving a boundary-value problem over the cross section, which requires finding the cross-sectional warping caused by

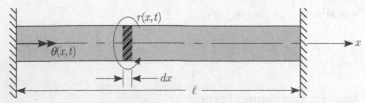

Figure 2.3. Beam undergoing torsional deformation

torsion. Although analytical solutions for this problem are available for simple cross-sectional geometries, solving for the cross-sectional warping and torsional stiffness is, in general, not a trivial exercise and possibly requires a numerical solution of Laplace's equation over the cross section. Moreover, when the beam is inhomogeneous with more than one constituent material and/or when one or more of the constituent materials is anisotropic, we must solve a more involved boundary-value problem over the cross-sectional area. For additional discussion of this point, see Section 2.4.

Equation of Motion. The beam is considered initially to have nonuniform properties along the x axis and to be loaded with a known, distributed twisting moment $r(x, t)$. The elastic twisting deflection, θ, is positive in a right-handed sense about this axis, as illustrated in Fig. 2.3. In contrast, the twisting moment, denoted by T, is the structural torque (i.e., the resultant moment of the tractions on a cross-sectional face about the elastic axis). Recall that an outward-directed normal on the positive x face is directed to the right, whereas an outward-directed normal on the negative x face is directed to the left. Thus, a positive torque tends to rotate the positive x face in a direction that is positive along the x axis in the right-hand sense and the negative x face in a direction that is positive along the $-x$ axis in the right-hand sense, as depicted in Fig. 2.3. This affects the boundary conditions, which are discussed in connection with applications of the theory in Chapter 3.

Letting $\overline{\rho I_p} dx$ be the polar mass moment of inertia about the x axis of the differential beam segment in Fig. 2.4, we can obtain the equation of motion by equating the resultant twisting moment on both segment faces to the rate of change of the segment's angular momentum about the elastic axis. This yields

$$T + \frac{\partial T}{\partial x} dx - T + r(x, t) dx = \overline{\rho I_p} dx \frac{\partial^2 \theta}{\partial t^2} \tag{2.39}$$

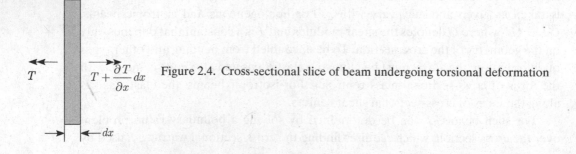

Figure 2.4. Cross-sectional slice of beam undergoing torsional deformation

or

$$\frac{\partial T}{\partial x} + r(x,t) = \overline{\rho I_p}\frac{\partial^2\theta}{\partial t^2} \tag{2.40}$$

where the polar mass moment of inertia is

$$\overline{\rho I_p} = \iint_{\mathcal{A}} \rho\left(y^2 + z^2\right) d\mathcal{A} \tag{2.41}$$

Here, $\mathcal{A}$ is the cross section of the beam, y and z are cross-sectional Cartesian coordinates, and ρ is the mass density of the beam. When ρ is constant over the cross section, then $\overline{\rho I_p} = \rho I_p$, where I_p is the polar area moment of inertia per unit length. In general, however, $\overline{\rho I_p}$ may vary along the x axis.

The twisting moment can be written in terms of the twist rate and the St. Venant torsional rigidity $\overline{GJ}$ as

$$T = \overline{GJ}\frac{\partial\theta}{\partial x} \tag{2.42}$$

Substituting these expressions into Eq. (2.40), we obtain the partial-differential equation of motion for the nonuniform beam given by

$$\frac{\partial}{\partial x}\left(\overline{GJ}\frac{\partial\theta}{\partial x}\right) + r(x,t) = \overline{\rho I_p}\frac{\partial^2\theta}{\partial t^2} \tag{2.43}$$

Strain Energy. The strain energy of an isotropic beam undergoing pure torsional deformation can be written as

$$U = \frac{1}{2}\int_0^\ell \overline{GJ}\left(\frac{\partial\theta}{\partial x}\right)^2 dx \tag{2.44}$$

This is also an appropriate expression of torsional strain energy for a composite beam without elastic coupling.

Kinetic Energy. The kinetic energy of a beam undergoing pure torsional deformation can be written as

$$K = \frac{1}{2}\int_0^\ell \overline{\rho I_p}\left(\frac{\partial\theta}{\partial t}\right)^2 dx \tag{2.45}$$

Virtual Work of Applied, Distributed Torque. The virtual work of an applied distributed twisting moment $r(x,t)$ on a beam undergoing torsional deformation may be computed as

$$\overline{\delta W} = \int_0^\ell r(x,t)\delta\theta(x,t)dx \tag{2.46}$$

where $\delta\theta$ is the variation of $\theta(x,t)$, the angle of rotation caused by twist. Note that $\delta\theta$ may be thought of as an increment of $\theta(x,t)$ that satisfies all geometric constraints.

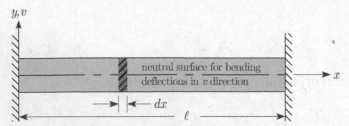

Figure 2.5. Schematic of beam for bending dynamics

2.3.2 Bending

As in the case of torsion, the beam is initially treated as having nonuniform properties along the x axis. The x axis is taken as the line of the individual cross-sectional neutral axes associated with pure bending in and normal to the plane of the diagram in Fig. 2.5. For simplicity, however, we consider only uncoupled bending in the x-y plane, thus excluding initially twisted beams from the development. The bending deflections are denoted by $v(x, t)$ in the y direction. The x axis is presumed to be straight, thus excluding initially curved beams. We continue to assume for now that the properties of the beam allow the x axis to be chosen so that bending and torsion are both structurally and inertially uncoupled. Therefore, in the plane(s) in which bending is taking place, the loci of both shear centers and mass centers must also coincide with the x axis. Finally, the transverse beam displacement, v, is presumed small to permit a linearly elastic representation of the deformation.

Equation of Motion. A free-body diagram for the differential-beam segment shown in Fig. 2.6 includes the shear force, V, and the bending moment, M. Recall from our previous discussion on torsion that an outward-directed normal on the positive x face is directed to the right, and an outward-directed normal is directed to the left on the negative x face. By this convention, V is the resultant of the transverse shear stresses

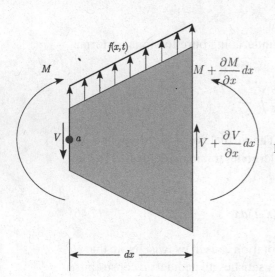

Figure 2.6. Schematic of differential beam segment

in the positive y direction (upward in Fig. 2.6) on a positive x cross-sectional face and in the negative y direction on a negative x cross-sectional face. In other words, a positive shear force tends to displace the positive x face upward and the negative x face downward, as depicted in Fig. 2.6. The bending moment, M, is the moment of the longitudinal stresses about a line parallel to the z axis (perpendicular to the plane of the diagram in Fig. 2.6) at the intersection between the cross-sectional plane and the neutral surface. Thus, a positive bending moment tends to rotate the positive x face positively about the z axis (in the right-handed sense) and the negative x face negatively about the z axis. This affects the boundary conditions, which are examined in detail in connection with applications of the theory in Chapter 3. The distributed loading (with units of force per unit length) is denoted by $f(x, t)$. The equation of motion for transverse-beam displacements can be obtained by setting the resultant force on the segment equal to the mass times the acceleration, which yields

$$f(x, t)dx - V + \left(V + \frac{\partial V}{\partial x}dx \right) = m\,dx\frac{\partial^2 v}{\partial t^2} \tag{2.47}$$

and leads to

$$\frac{\partial V}{\partial x} + f(x, t) = m\frac{\partial^2 v}{\partial t^2} \tag{2.48}$$

where m is the mass per unit length, given by ρA for homogeneous cross sections. We must also consider the moment equation. We note here that the cross-sectional rotational inertia about the z axis will be ignored because it has a small effect. Taking a counterclockwise moment as positive, we sum the moments about the point a to obtain

$$-M + \left(M + \frac{\partial M}{\partial x}dx \right) + \left(V + \frac{\partial V}{\partial x}dx \right)dx + \left(f - m\frac{\partial^2 v}{\partial t^2} \right)\frac{(dx)^2}{2} = 0 \tag{2.49}$$

which, after we neglect the higher-order differentials (i.e., higher powers of dx), becomes

$$\frac{\partial M}{\partial x} + V = 0 \tag{2.50}$$

Recall that the bending moment is proportional to the local curvature; therefore

$$M = \overline{EI}\frac{\partial^2 v}{\partial x^2} \tag{2.51}$$

where $\overline{EI}$ may be regarded as the effective bending stiffness of the beam at a particular cross section and hence may vary with x. Note that for isotropic beams, calculation of the bending rigidity is a straightforward integration over the cross section, given by

$$\overline{EI} = \iint_A Ey^2 d\mathcal{A} \tag{2.52}$$

where E is the Young's modulus. When the beam is homogeneous the Young's modulus may be moved outside the integration so that $\overline{EI} = EI$ where I is the cross-sectional area moment of inertia about the z axis for a particular cross section. Here, the origin of the y and z axes is at the sectional centroid. However, when one or more of the constituent materials is anisotropic, determination of the effective bending rigidity becomes more difficult to perform rigorously. For additional discussion of this point, see Section 2.4.

Substitution of Eq. (2.51) into Eq. (2.50) and of the resulting equation into Eq. (2.48) yields the partial differential equation of motion for a spanwise nonuniform beam as

$$\frac{\partial^2}{\partial x^2}\left(\overline{EI}\frac{\partial^2 v}{\partial x^2}\right) + m\frac{\partial^2 v}{\partial t^2} = f(x,t) \tag{2.53}$$

Strain Energy. The strain energy of an isotropic beam undergoing pure bending deformation can be written as

$$U = \frac{1}{2}\int_0^\ell \overline{EI}\left(\frac{\partial^2 v}{\partial x^2}\right)^2 dx \tag{2.54}$$

This is also an appropriate expression for the bending strain energy for a composite beam without elastic coupling.

Kinetic Energy. The kinetic energy of a beam undergoing bending deformation can be written as

$$K = \frac{1}{2}\int_0^\ell m\left(\frac{\partial v}{\partial t}\right)^2 dx \tag{2.55}$$

just as for a vibrating string. For a spanwise nonuniform beam, m may vary with x.

Virtual Work of Applied, Distributed Force. The virtual work of an applied distributed force $f(x,t)$ on a beam undergoing bending deformation may be computed as

$$\overline{\delta W} = \int_0^\ell f(x,t)\delta v(x,t)dx \tag{2.56}$$

just as for a vibrating string.

2.4 Composite Beams

Recall that the x axis (i.e., the axial coordinate) for homogeneous, isotropic beams is generally chosen as the locus of cross-sectional shear centers. This choice is frequently denoted as the "elastic axis" because it structurally uncouples torsion from both transverse shearing deformation and bending. Thus, transverse forces acting through this axis do not twist the beam. However, even for spanwise uniform composite beams, when transverse shear forces act through any axis defined as the locus of a cross-sectional property, it is still possible that these forces will twist the beam because bending-twist coupling may be present. For the type of composite-beam analysis presented herein, we still choose the x axis to be along the locus of shear centers; but, for composite beams, this choice uncouples only torsion and transverse shear deformation. Therefore, although transverse shear forces acting through the x axis do not *directly* induces twist, the bending moment induced by the shear force still induces twist when bending-twist coupling is present.

2.4.1 Constitutive Law and Strain Energy for Coupled Bending and Torsion

A straightforward way to introduce such coupling in the elementary beam equations presented previously is to alter the "constitutive law" (i.e., the relationship between cross-sectional stress resultants and the generalized strains). So, we change

$$
\begin{Bmatrix} T \\ M \end{Bmatrix} = \begin{bmatrix} \overline{GJ} & 0 \\ 0 & \overline{EI} \end{bmatrix} \begin{Bmatrix} \dfrac{\partial \theta}{\partial x} \\ \dfrac{\partial^2 v}{\partial x^2} \end{Bmatrix}
\tag{2.57}
$$

to

$$
\begin{Bmatrix} T \\ M \end{Bmatrix} = \begin{bmatrix} \overline{GJ} & -K \\ -K & \overline{EI} \end{bmatrix} \begin{Bmatrix} \dfrac{\partial \theta}{\partial x} \\ \dfrac{\partial^2 v}{\partial x^2} \end{Bmatrix}
\tag{2.58}
$$

where $\overline{EI}$ is the effective bending stiffness, $\overline{GJ}$ is the effective torsional stiffness, and K is the effective bending-torsion coupling stiffness (with the same dimensions as $\overline{EI}$ and $\overline{GJ}$). Whereas $\overline{EI}$ and $\overline{GJ}$ are strictly positive, K may be positive, negative, or zero. A positive value of K implies that when the beam is loaded with an upward vertical force at the tip, the resulting positive bending moment induces a positive (i.e., nose-up) twisting moment. Values of $\overline{GJ}, \overline{EI}$, and K are best found by cross-sectional codes such as VABS,™ a commercially available computer program developed at Georgia Tech (Hodges, 2006).

Now, given Eq. (2.58), it is straightforward to write the strain energy as

$$
U = \frac{1}{2} \int_0^\ell \begin{Bmatrix} \dfrac{\partial \theta}{\partial x} \\ \dfrac{\partial^2 v}{\partial x^2} \end{Bmatrix}^T \begin{bmatrix} \overline{GJ} & -K \\ -K & \overline{EI} \end{bmatrix} \begin{Bmatrix} \dfrac{\partial \theta}{\partial x} \\ \dfrac{\partial^2 v}{\partial x^2} \end{Bmatrix} dx
\tag{2.59}
$$

where the 2×2 matrix must be positive-definite for physical reasons. This means that of necessity, $K^2 < \overline{EI}\,\overline{GJ}$.

2.4.2 Inertia Forces and Kinetic Energy for Coupled Bending and Torsion

In general, there is also inertial coupling between bending and torsional deflections. This type of coupling stems from d, the offset of the cross-sectional mass centroid from the x axis shown in Fig. 2.7 and given by

$$
md = -\iint_A \rho \, z \, d\mathcal{A}
\tag{2.60}
$$

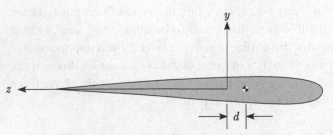

Figure 2.7. Cross section of beam for coupled bending and torsion

so that the acceleration of the mass centroid is

$$\mathbf{a}_C = \left(\frac{\partial^2 v}{\partial t^2} + d\frac{\partial^2 \theta}{\partial t^2}\right)\hat{\mathbf{j}} \tag{2.61}$$

For inhomogeneous beams, the offset d may be defined as the distance from the x axis to the cross-sectional mass centroid, positive when the mass centroid is toward the leading edge from the x axis.

Similarly, if one neglects rotary inertia of the cross-sectional plane about the z axis, the angular momentum of the cross section about P is

$$\mathbf{H}_P = \left(\overline{\rho I_p}\frac{\partial \theta}{\partial t} + md\frac{\partial v}{\partial t}\right)\hat{\mathbf{i}} \tag{2.62}$$

where, for beams in which the material density varies over the cross section, we may calculate the cross-sectional mass polar moment of inertia as

$$\overline{\rho I_p} = \iint\limits_{\mathcal{A}} \rho\left(y^2 + z^2\right) d\mathcal{A} \tag{2.63}$$

The kinetic energy follows from similar considerations and can be written directly as

$$K = \frac{1}{2}\int_0^\ell \left[m\left(\frac{\partial v}{\partial t}\right)^2 + 2md\frac{\partial \theta}{\partial t}\frac{\partial v}{\partial t} + \overline{\rho I_p}\left(\frac{\partial \theta}{\partial t}\right)^2 \right] dx \tag{2.64}$$

2.4.3 Equations of Motion for Coupled Bending and Torsion

Using the coupled constitutive law and inertia forces, the partial differential equations of motion for coupled bending and torsion of a composite beam become

$$\overline{\rho I_p}\frac{\partial^2 \theta}{\partial t^2} + md\frac{\partial^2 v}{\partial t^2} - \frac{\partial}{\partial x}\left(\overline{GJ}\frac{\partial \theta}{\partial x} - K\frac{\partial^2 v}{\partial x^2}\right) = r(x,t)$$

$$m\left(\frac{\partial^2 v}{\partial t^2} + d\frac{\partial^2 \theta}{\partial t^2}\right) + \frac{\partial^2}{\partial x^2}\left(\overline{EI}\frac{\partial^2 v}{\partial x^2} - K\frac{\partial \theta}{\partial x}\right) = f(x,t) \tag{2.65}$$

where we see the structural coupling through K and the inertial coupling through d.

Of course, we may simplify these equations for isotropic beams undergoing coupled bending and torsion simply by setting $K = 0$.

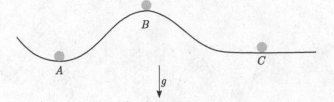

Figure 2.8. Character of static-equilibrium positions

2.5 The Notion of Stability

Consider a structure undergoing external loads applied quasistatically. In such a case, static equilibrium is maintained as the elastic structure deforms. If now at any level of the external force a "small" external disturbance is applied, and the structure reacts by simply performing oscillations about the deformed equilibrium state, the equilibrium state is said to be stable. This disturbance can be in the form of deformation or velocity; by "small," we mean "as small as desired." As a result of this latter definition, it is more appropriate to say that the equilibrium is stable for a small disturbance. In addition, we stipulate that when the disturbance is introduced, the level of the external forces is kept constant. Conversely, if the elastic structure either (a) tends to and remains in the disturbed position, or (b) diverges from the equilibrium state, the equilibrium is said to be unstable. Some authors prefer to distinguish these two conditions and call the equilibrium "neutrally stable" for case (a) and "unstable" for case (b). When either of these two cases occurs, the level of the external forces is referred to as "critical."

This is illustrated using the system shown in Fig. 2.8. This system consists of a ball of weight W resting at different points on a surface with zero curvature normal to the plane of the figure. Points of zero slope on the surface denote positions of static equilibrium (i.e., points A, B, and C). Furthermore, the character of equilibrium at these points is substantially different. At A, if the system is disturbed through infinitesimal disturbances (i.e., small displacements or small velocities), it simply oscillates, about the static-equilibrium position A. Such an equilibrium position is called stable for small disturbances. At point B, if the system is disturbed, it tends to move away from the static-equilibrium position B. Such an equilibrium position is called unstable for small disturbances. Finally, at point C, if the system is disturbed, it tends to remain in the disturbed position. Such an equilibrium position is called neutrally stable or indifferent for small disturbances. The expression "for small disturbances" is used because the definition depends on the small size of the perturbations and is the foundational reason we may use linearized equations to conduct the analysis. If the disturbances are allowed to be of finite magnitude, then it is possible for a system to be unstable for small disturbances but stable for large disturbances (i.e., point B, Fig. 2.9a) or stable for small disturbances but unstable for large disturbances (i.e., point A, Fig. 2.9b).[1]

---

[1] Portions of section 2.5 including figures 2.8 and 2.9 are excerpted from Simitses and Hodges (2006) and (2010), used with permission.

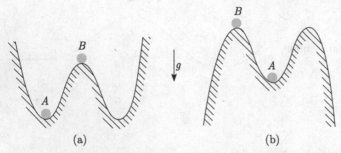

Figure 2.9. Character of static-equilibrium positions for finite disturbances

2.6 Systems with One Degree of Freedom

The behavior of systems with one degree of freedom is of interest in its own right. Through the various forms of modal approximations, such as the Ritz and Galerkin methods (see Section 3.5), the behavior of complex systems frequently can be reduced to a set of equations each having the form of a single-degree-of-freedom system. Therefore, it is worthwhile to explore the various types of behavior we associate with such systems.

Consider a particle of mass m, restrained by a spring with elastic constant k and a damper with damping constant c, and forced with a function $f(t)$ (Fig. 2.10). The governing equation can be written

$$m\ddot{x} + c\dot{x} + kx = f(t) \tag{2.66}$$

where $x(t)$ is the single unknown, typically a displacement or rotation but not limited to such. Here, the mass m, the damping c, and the stiffness k are the system parameters and $f(t)$ is a forcing function. Our interest in this system is limited for now in two special cases: (1) unforced motion, with $f(t)$ as identically zero; and (2) harmonically forced motion.

2.6.1 Unforced Motion

Eq. (2.66) for unforced motion is given by

$$m\ddot{x} + c\dot{x} + kx = 0 \tag{2.67}$$

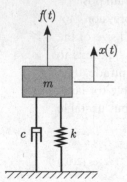

Figure 2.10. Single-degree-of-freedom system

An exhaustive treatment of this equation is beyond the scope of this text. Suffice it to say that for our purposes, we are concerned with the qualitative aspects of the motion for various combinations of parameter values. We are interested in both positive and negative stiffness and damping. To facilitate exploration of the behavior, we define the natural frequency ω such that $k = m\omega^2$, divide the equation by m, and introduce the damping ratio ζ so that $c = 2m\zeta\omega$. Then, the equation of motion reduces to

$$\ddot{x} + 2\zeta\omega\dot{x} + \omega^2 x = 0 \tag{2.68}$$

another advantageous step is to introduce dimensionless time $\psi = \omega t$, derivatives with respect to which are denoted by $(\ )'$. With this, Eq. (2.68) becomes

$$x'' + 2\zeta x' + x = 0 \tag{2.69}$$

We are mostly concerned about the response of systems with small damping ratios, in which $\zeta < 1$. For this case, the general solution is

$$x(\psi) = e^{-\zeta\psi}\left[a\cos\left(\sqrt{1-\zeta^2}\psi\right) + b\sin\left(\sqrt{1-\zeta^2}\psi\right)\right]$$
$$x'(\psi) = e^{-\zeta\psi}\left[(b\sqrt{1-\zeta^2} - \zeta a)\cos\left(\sqrt{1-\zeta^2}\psi\right)\right. \tag{2.70}$$
$$\left. - (\zeta b + a\sqrt{1-\zeta^2})\sin\left(\sqrt{1-\zeta^2}\psi\right)\right]$$

The responses caused by arbitrary initial displacement and velocity can be constructed by combining the responses to unit initial displacement and unit initial velocity. For the first, we let $x(0) = 1$ and $x'(0) = 0$, which together imply that $a = 1$ and $b = \zeta/\sqrt{1-\zeta^2}$. For the second, we let $x(0) = 0$ and $x'(0) = 1$, which together imply $a = 0$ and $b = 1/\sqrt{1-\zeta^2}$. In all cases, the displacement and velocity both exhibit an oscillatory character with a decaying amplitude for $\zeta > 0$ and a growing amplitude for $\zeta < 0$. Positive damping leads to a decaying response signal (Fig. 2.11) and negative damping to a growing response signal (Fig. 2.12).

Actual mechanical systems always have positive stiffness. However, with the advent of active materials, it is possible to have a negative effective stiffness. Also, in the field of aeroelasticity, aerodynamics can contribute a negative stiffening effect that possibly overpowers the positive stiffness from the structure or the support. When a system has a negative effective stiffness, the response can be written as

$$x(\psi) = e^{-\zeta t}\left[a\cosh\left(t\sqrt{1+\zeta^2}\right) + b\sinh\left(t\sqrt{1+\zeta^2}\right)\right] \tag{2.71}$$

This function is only slightly affected by the sign of ζ and the initial conditions. Typical results are shown in Fig. 2.13. Response for negative stiffness is nonoscillatory divergent motion. Damping makes little difference when the stiffness is negative, although response is slightly larger for negative damping. In summary, when instabilities are encountered, the system response is divergent and may be either nonoscillatory or oscillatory.

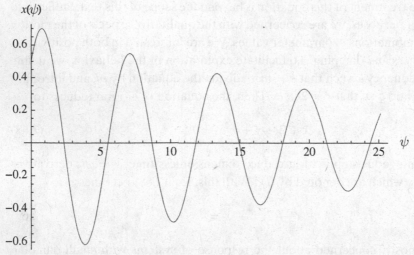

Figure 2.11. Response for system with positive k and $x(0) = x'(0) = 0.5$, $\zeta = 0.04$

2.6.2 Harmonically Forced Motion

We now consider the case of harmonically forced motion, where $f(t)$ is a harmonic function of the form $f(t) = kA\cos(\Omega t)$. The response to harmonically excited motion is a subject worthy of study, but we hardly "scratch the surface" in this brief discussion. For the present purpose, we consider the equation of motion written as

$$m\ddot{x} + c\dot{x} + kx = kA\cos(\Omega t) \tag{2.72}$$

Dividing through by m, we find

$$\ddot{x} + 2\zeta\omega\dot{x} + \omega^2 x = A\omega^2\cos(\Omega t) = A\omega^2 e^{i\Omega t} \tag{2.73}$$

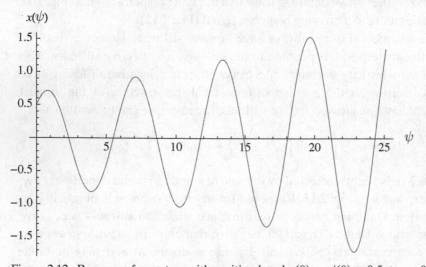

Figure 2.12. Response for system with positive k and $x(0) = x'(0) = 0.5$, $\zeta = -0.04$

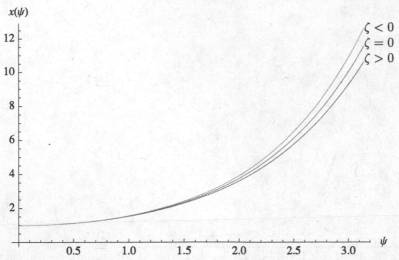

Figure 2.13. Response for system with negative k and $x(0) = 1$, $x'(0) = 0$, $\zeta = -0.05$, 0, and 0.05.

with X as a complex variable and the actual displacement being found as the real part of x. Considering only the steady-state part of the response, we may assume

$$x = Xe^{i\Omega t} \tag{2.74}$$

Substitution of Eq. (2.74) into Eq. (2.73) yields

$$\frac{X}{A} = G(i\Omega) = \frac{1}{1 - \left(\frac{\Omega}{\omega}\right)^2 + 2i\zeta\frac{\Omega}{\omega}} \tag{2.75}$$

where $G(i\Omega)$ is the frequency response. This form allows us to write the solution as

$$x = A|G(i\Omega)|\cos(\Omega t - \phi) \tag{2.76}$$

where $|G(i\Omega)|$ is known as the magnification factor, given by

$$|G(i\Omega)| = \frac{1}{\sqrt{4\left(\frac{\Omega}{\omega}\right)^2\zeta^2 + \left[1 - \left(\frac{\Omega}{\omega}\right)^2\right]^2}} \tag{2.77}$$

and plotted in Fig. 2.14. The phase angle

$$\phi = \tan^{-1}\left(\frac{2\zeta\frac{\Omega}{\omega}}{1 - \left(\frac{\Omega}{\omega}\right)^2}\right) \tag{2.78}$$

shows the delay between the peaks in $f(t)$ at $t = 2n\pi/\Omega$, and the peaks in $x(t)$ at $t = \phi/\Omega + 2n\pi/\Omega$ for $n = 0, 1, \ldots$.

Now, as an example, we may consider a harmonic forcing function $f(t)$ with $A = 1$ plotted along with $x(t)$ for a particular choice ζ and Ω/ω in Fig. 2.15. Here, the phase lag is noticeable as the response peaks are shifted approximately 43.45 degrees to the right.

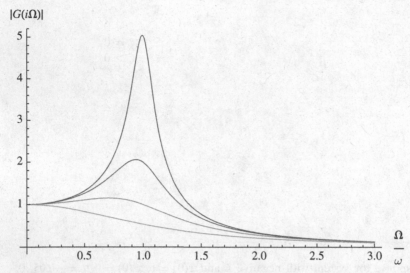

Figure 2.14. Magnification factor $|G(i\Omega)|$ versus Ω/ω for various values of ζ for a harmonically excited system

Harmonically forced systems also may exhibit a large and possibly dangerous response in the case of resonance, where the driving frequency Ω is very near ω. For undamped systems, the response grows as

$$\frac{A}{2}\left[t\sin(\omega t) + \cos(\omega t)\right] \tag{2.79}$$

whereas the response amplitude reaches $1/(2\zeta)$ for lightly damped systems.

2.7 Epilogue

In this chapter, we laid out the foundational theories of mechanics that are needed for an introductory treatment of structural dynamics and aeroelasticity. It is hoped

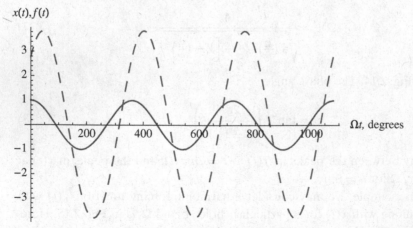

Figure 2.15. Excitation $f(t)$ (solid line) and response $x(t)$ (dashed line) versus Ωt (in degrees) for $\zeta = 0.1$ and $\Omega/\omega = 0.9$ for a harmonically excited system

that students find it helpful to be able to refer to these treatments as they are applied throughout the remainder of the text. Structural dynamics and aeroelasticity analyses of realistic aircraft structural elements may require more sophisticated theories, such as plate and shell theory and full three-dimensional finite-element analysis; however, such treatments are beyond the scope of this text.

Problems

1. Show that the equation of motion for longitudinal vibration of a uniform beam is the same as that for a string, viz.

$$EA\frac{\partial^2 u}{\partial x^2} = m\frac{\partial^2 u}{\partial t^2}$$

2. Show that the strain energy for longitudinal deformation of a beam is the same as that for a string, viz.

$$P = \int_0^\ell EA\left(\frac{\partial u}{\partial x}\right)^2 dx$$

3. Show that the kinetic energy for longitudinal deformation of a beam is the same as that for a string, viz.

$$K = \int_0^\ell m\left(\frac{\partial u}{\partial t}\right)^2 dx$$

4. Show that Eqs. (2.65) are the equations of motion for coupled bending-torsion behavior of a composite beam.

3 Structural Dynamics

O students, study mathematics, and do not build without foundations....
—Leonardo da Vinci

The purpose of this chapter is to convey to students a small introductory portion of the theory of structural dynamics. Much of the theory to which the students will be exposed in this treatment was developed by mathematicians during the time between Newton and Rayleigh. The grasp of this mathematical foundation is therefore a goal that is worthwhile in its own right. Moreover, as implied by the da Vinci quotation, a proper use of this foundation enables the advance of technology.

Structural dynamics is a broad subject, encompassing determination of natural frequencies and mode shapes (i.e., the so-called free-vibration problem), response due to initial conditions, forced response in the time domain, and frequency response. In the following discussion, we deal with all except the last category. For response problems, if the loading is at least in part of aerodynamic origin, then the response is said to be aeroelastic. In general, the aerodynamic loading then will depend on the structural deformation, and the deformation will depend on the aerodynamic loading. Linear aeroelastic problems are considered in subsequent chapters, and linear structured dynamics problems are considered in the present chapter. Other important phenomena, such as limit-cycle oscillations of lifting surfaces, must be treated with sophisticated nonlinear-analysis methodology; however, they are beyond the scope of this text.

The value of structural dynamics to the general study of aeroelastic phenomena is its ability to provide a means of quantitatively describing the deformation pattern at any instant in time for a continuous structural system in response to external loading. Although there are many methods of approximating the structural-deformation pattern, several of the widely used methods are reducible to what is called a "modal representation" as long as the underlying structural modeling is linear. The purpose of this chapter is to establish the concept of modal representation and show how it can be used to describe the dynamic behavior of continuous elastic systems. Also included is an introductory treatment of the Ritz and Galerkin methods, techniques that use assumed modes or similar sets of functions to obtain approximate solutions in a simple way. Indeed, both methods are close relatives of the finite element method, a widely used approximate method that can accurately analyze realistic structural configurations. Only the basics of applying the finite element method to beams are covered herein; details of this method are in books that offer a more advanced perspective on structural analysis, several of which are listed in the references.

The analytical developments presented in this chapter are conceptually similar to the methods of analysis conducted on complete flight vehicles. In an effort to maintain analytical simplicity, the continuous structural configurations to be examined are all uniform and one-dimensional. Although such structures may appear impractical relative to conventional aircraft, they exhibit structural dynamic properties and representations that are essentially the same as those of full-scale flight vehicles.

3.1 Uniform String Dynamics

To more easily understand a mathematical description of the mechanics associated with the structural dynamics of continuous elastic systems, the classical "vibrating-string problem" is first considered. Although the free-vibration of a string can be described by the linear second-order partial differential equation in one dimension derived in Chapter 2 (see Eq. 2.26), it is typically descriptive of the more complex linearly elastic systems of aerospace vehicles. After the fundamental concepts are reviewed for the string, other components that are more representative of these vehicles are discussed. Although the free vibration of a string can be analyzed using equations of motion of the same form as those governing uniform beam extensional and torsional vibrations, the string is chosen as our first example primarily because—in contrast to the behavior of the other structures—string behavior can be visualized easily. Moreover, typically by this time in their undergraduate studies, most students have had some exposure to the solution of string-vibration problems.

3.1.1 Standing Wave (Modal) Solution

The wave equation governing transverse vibration of a nonuniform string was derived in Eq. (2.26) for uniform strings. Here, we repeat it for convenience, with a slight generalization

$$T\frac{\partial^2 v}{\partial x^2} = m(x)\frac{\partial^2 v}{\partial t^2} \tag{3.1}$$

Here, the mass distribution $m(x)$ is allowed to vary along the string. This partial differential equation of motion with two independent variables may be reduced to two ordinary differential equations by making a "separation of variables." The dependent variable of transverse displacement is represented by

$$v(x, t) = X(x)Y(t) \tag{3.2}$$

This product form is now substituted into the wave equation, Eq. (3.1). To simplify the notation, let $(\ )'$ and $(\dot{\ })$ denote ordinary derivatives with respect to x and t. Thus, the wave equation becomes

$$TX''(x)Y(t) = m(x)X(x)\ddot{Y}(t) \tag{3.3}$$

Rearranging terms as

$$\frac{TX''(x)}{m(x)X(x)} = \frac{\ddot{Y}(t)}{Y(t)} \tag{3.4}$$

we observe that the left-hand side of this equation is a function of only the single independent variable x and the right-hand side is a function of only t. The presence of $m(x)$ reflects material density and/or geometry that varies along the string, whereas constant T is consistent with approximations used in deriving Eq. (3.1). Because each side of the equation is a function of different independent variables, the only way that the equality can be valid is for each side to be equal to a common constant. Let this constant be $-\omega^2$, so that

$$\frac{TX''(x)}{m(x)X(x)} = \frac{\ddot{Y}(t)}{Y(t)} = -\omega^2 \tag{3.5}$$

This yields two ordinary differential equations, given by

$$TX''(x) + m(x)\omega^2 X(x) = 0$$
$$\ddot{Y}(t) + \omega^2 Y(t) = 0 \tag{3.6}$$

Both of these equations are linear, ordinary differential equations. The second of the two equations has constant coefficients and is the governing equation for a harmonic oscillator with frequency ω. Because the first equation has a variable coefficient $m(x)$ in its second term, however, it can be solved in closed form only in special cases.

Specifically, when the mass per unit length m is a constant, the first of Eqs. (3.6) has a familiar solution. In this case, it is expedient to introduce

$$\alpha^2 = \frac{m\omega^2}{T} \tag{3.7}$$

so that the two ordinary differential equations are of the same form; that is

$$X''(x) + \alpha^2 X(x) = 0$$
$$\ddot{Y}(t) + \omega^2 Y(t) = 0 \tag{3.8}$$

Because the general solutions to these linear, second-order differential equations are well known, they are written without any further justification as

$$X(x) = A\sin(\alpha x) + B\cos(\alpha x)$$
$$Y(t) = C\sin(\omega t) + D\cos(\omega t) \tag{3.9}$$

where

$$\omega = \alpha\sqrt{\frac{T}{m}} \tag{3.10}$$

Recall that these solutions are only valid when $\alpha \neq 0$.

The boundary conditions on the string are given in Eqs. (2.27). The boundary condition on the left end of the string, where $x = 0$, can be written as

$$v(0, t) = X(0)Y(t) = 0 \qquad (3.11)$$

which is satisfied when

$$X(0) = 0 \qquad (3.12)$$

so that

$$B = 0 \qquad (3.13)$$

The boundary condition on the right end is

$$v(\ell, t) = X(\ell)Y(t) = 0 \qquad (3.14)$$

which is satisfied when

$$X(\ell) = 0 \qquad (3.15)$$

and so

$$A\sin(\alpha\ell) = 0 \qquad (3.16)$$

If $A = 0$, the displacement is identically zero for all x and t. Although this is an acceptable solution, it is of little interest and therefore is referred to as a "trivial solution." Of more concern is when

$$\sin(\alpha\ell) = 0 \qquad (3.17)$$

This relationship is called the "characteristic equation" and has a denumerably infinite set of solutions known as "eigenvalues." These solutions can be written as

$$\alpha_i = \frac{i\pi}{\ell} \quad (i = 1, 2, \ldots) \qquad (3.18)$$

where, recalling that $\alpha \neq 0$ is a requirement for this solution, we must exclude the root corresponding to $i = 0$. To ascertain whether a nontrivial $\alpha = 0$ solution exists, we must return to the first of Eqs. (3.8) and determine whether a nontrivial solution exists with $\alpha = 0$ that also satisfies all the boundary conditions—that is, whether there is a nontrivial solution to $X'' = 0$ for which $X(0) = X(\ell) = 0$. Obviously, there is no such solution for this problem. Solutions associated with $\alpha = 0$ are addressed in more detail when we consider problems for which rigid-body modes may exist.

Therefore, for each integer value of the index i, there is an eigenvalue α_i and an associated solution X_i, called the "eigenfunction." It contributes to the general solution based on the corresponding value of Y_i. Thus, its total contribution can be written as

$$v_i(x, t) = X_i(x)Y_i(t) \qquad (3.19)$$

where

$$X_i(x) = A_i \sin(\alpha_i x)$$

$$Y_i(t) = C_i \sin(\omega_i t) + D_i \cos(\omega_i t)$$

(3.20)

The constants A_i, C_i, and D_i may have different numerical values for each eigenvalue; thus, they are subscripted with the index. The most general solution for the string displacement would have contributions associated with all the eigenvalues. Thus, the general solution can be written as a sum of the complete set as

$$v(x, t) = \sum_{i=1}^{\infty} v_i(x, t)$$

$$= \sum_{i=1}^{\infty} \sin\left(\frac{i\pi x}{\ell}\right) [E_i \sin(\omega_i t) + F_i \cos(\omega_i t)]$$

(3.21)

where

$$\omega_i = \frac{i\pi}{\ell}\sqrt{\frac{T}{m}}$$

(3.22)

Note that the original constants were combined as

$$E_i = A_i C_i$$

$$F_i = A_i D_i$$

(3.23)

Close inspection of this total string displacement indicates that at any given instant, the transverse deflection is represented by summation over a denumerably infinite set of shapes. Each shape is of indeterminate amplitude and is associated with a particular eigenfunction; these shapes are also called "mode shapes" in the field of structural dynamics. They are represented here by $\phi_i(x)$. Thus, for transverse deflection of a string, the mode shapes may be written as

$$\phi_i(x) = \sin\left(\frac{i\pi x}{\ell}\right)$$

(3.24)

or any constant times $\phi_i(x)$. It can be observed from this function (Fig. 3.1) that the higher the mode number i, the more crossings of the zero axis on the interval $0 < x < \ell$. These crossings are sometimes referred to as "nodes." The trend of increasing numbers of nodes with an increase in the mode number is generally true in structural dynamics.

In the previous solution for the total displacement, each mode shape is multiplied by a function of time. This multiplier is called the "generalized coordinate" and is represented here by $\xi_i(t)$. For this specific problem, the generalized coordinates are

$$\xi_i(t) = E_i \sin(\omega_i t) + F_i \cos(\omega_i t)$$

(3.25)

and thus are seen to be simple harmonic functions of time with frequencies ω_i. Because there were no external loads applied to the string, the preceding result is

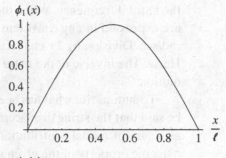

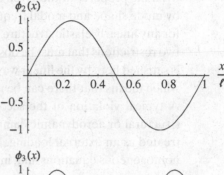

Figure 3.1. First three mode shapes for vibrating string

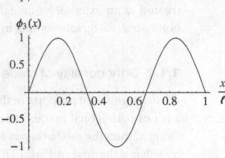

called the "homogeneous solution." Had there been an external loading, the resulting time dependency of the generalized coordinates would reflect it.

Thus, the total string displacement can be written as a sum of "modal" contributions of the form

$$v(x,t) = \sum_{i=1}^{\infty} \phi_i(x)\xi_i(t) \tag{3.26}$$

This expression can be interpreted as a weighted sum of the mode shapes, each of which has a modal amplitude (i.e., the generalized coordinate) that is a function of time. For the homogeneous solution obtained here, this time dependency is simple harmonic at a frequency that is unique for each mode or eigenvalue. These are called the "natural frequencies" of the modes, or "modal frequencies," and are represented by ω_i. For the string, they are

$$\omega_i = \frac{i\pi}{\ell}\sqrt{\frac{T}{m}} \quad i = 1, 2, \ldots, \infty \tag{3.27}$$

with the lowest frequencies given by the lowest mode numbers. Indeed, just as the increase in the number of nodes with the mode number is generally true, so it is with the natural frequency. When the physical and geometric parameters of the problem are expressed in any consistent[1] set of units, the units of the natural frequency are rad/sec. Division by 2π converts the units of frequency into "cycles per second," or Hertz. The inverse of the natural frequency in Hertz is the period of the oscillatory motion.

To summarize what has been accomplished in solving the wave equation, it may be said that the string displacement as a function of both x and t can be represented as a sum of modal contributions. Each mode in this representation is a structural dynamic property of the given system (i.e., string) and can be described completely by mode shape and modal frequency. Such "modes of vibration" can be formulated for any linearly elastic structure that is a conservative system. This statement includes two restrictions that must be observed for a modal representation: (1) linearity, which is satisfied here by the linear wave equation; and (2) the system must be conservative, which means that there can be no addition or dissipation of energy in free vibration. A typical violation of the second restriction is the existence of damping, such as structural or aerodynamic damping. When damping is present, it can be adequately treated as an external loading. Mode shapes are determined only by the solution of homogeneous equations and, in general, they are real only for self-adjoint equations.

3.1.2 Orthogonality of Mode Shapes

A most significant property of the mode shapes derived for the string is that they form a set of orthogonal mathematical functions. If the mass distribution is nonuniform along x, then the mode shapes are no longer $\sin(i\pi x/\ell)$; instead, they must be found by solving the first of Eqs. (3.6). The resulting mode shapes, however, may not be expressible in closed form. Nonetheless, they are orthogonal but with respect to the mass distribution as a weighting function. In such a case, this condition of functional orthogonality can be described analytically as

$$\int_0^\ell m(x)\phi_i(x)\phi_j(x)\,dx = 0 \quad (i \neq j)$$

$$\neq 0 \quad (i = j)$$

(3.28)

To prove that the mode shapes obtained for the string problem are orthogonal, an individual modal contribution given by

$$v_i(x, t) = \phi_i(x)\xi_i(t)$$

(3.29)

where $\phi_i(x)$ is a normalized solution of the first of Eqs. (3.6). Substituting $v_i(x, t)$ into the governing differential equation (i.e., wave equation), we obtain

$$T\frac{\partial^2 v_i}{\partial x^2} = m\frac{\partial^2 v_i}{\partial t^2}$$

(3.30)

---

[1] For example, with SI units, one has the units of T as N, m as kg/m, and ℓ as m. With English units, one has the units of T as lb, m as slugs/ft, and ℓ as ft.

or

$$T\phi_i''(x)\xi_i(t) = m(x)\phi_i(x)\ddot{\xi}_i(t) \tag{3.31}$$

Because the general (i.e., homogeneous) solution for the generalized coordinate is a simple harmonic function, then we may write

$$\ddot{\xi}_i = -\omega_i^2\,\xi_i \tag{3.32}$$

Thus, the wave equation becomes

$$T\phi_i''(x)\xi_i(t) = -m(x)\phi_i(x)\omega_i^2\xi_i(t) \tag{3.33}$$

so that

$$T\phi_i''(x) = -m(x)\phi_i(x)\omega_i^2 \tag{3.34}$$

If this procedure is repeated by substituting the jth modal contribution into the wave equation, a similar result

$$T\phi_j''(x) = -m(x)\phi_j(x)\omega_j^2 \tag{3.35}$$

is obtained. After multiplying Eq. (3.34) by ϕ_j and Eq. (3.35) by ϕ_i, subtracting, and integrating the result over the length of the string, we obtain

$$(\omega_i^2 - \omega_j^2)\int_0^\ell m(x)\phi_i(x)\phi_j(x)dx = T\int_0^\ell \left[\phi_i(x)\phi_j''(x) - \phi_i''(x)\phi_j(x)\right]dx \tag{3.36}$$

The integral on the right-hand side can be integrated by parts using

$$\int_a^b u\,dv = uv\Big|_a^b - \int_a^b v\,du \tag{3.37}$$

by letting

$$\begin{aligned} u &= \phi_i & du &= \phi_i'dx \\ v &= \phi_j' & dv &= \phi_j''dx \end{aligned} \tag{3.38}$$

for the first term and

$$\begin{aligned} u &= \phi_j & du &= \phi_j'dx \\ v &= \phi_i' & dv &= \phi_i''dx \end{aligned} \tag{3.39}$$

for the second. The result becomes

$$(\omega_i^2 - \omega_j^2)\int_0^\ell m(x)\phi_i(x)\phi_j(x)dx$$
$$= T\left(\phi_i\phi_j' - \phi_i'\phi_j\right)\Big|_0^\ell - T\int_0^\ell (\phi_i'\phi_j' - \phi_i'\phi_j')dx = 0 \tag{3.40}$$

Every term on the right-hand side is zero: the first and second because the mode shape is zero at both ends by virtue of the boundary conditions given by Eqs. (2.27), and the integral because of cancellation. It may now be concluded that when $i \neq j$, because $\omega_i \neq \omega_j$, it follows that

$$\int_0^\ell m(x)\phi_i(x)\phi_j(x)dx = 0 \tag{3.41}$$

This relationship thus demonstrates that the mode shapes for a string that is fixed at both ends form an orthogonal set of functions. However, when $i = j$

$$\int_0^\ell m(x)\phi_i^2(x)dx = M_i \tag{3.42}$$

The value of this integral, M_i, is referred to as the "generalized mass" of the ith mode. The numerical values of the generalized masses depend on the normalization scheme used for the mode shapes $\phi_i(x)$.

This development is for a string of nonuniform mass per unit length and constant tension force. It is important to note that it readily can be generalized to more involved developments for beam torsional and bending deformation. In such cases, the structural stiffnesses—which are analogous to the tension force in the string problem—also may be nonuniform along the span. Although the structural stiffnesses may not be taken outside the integrals in such cases, the rest of the development remains similar. See Problems 8(a) and 10(a) at the end of this chapter.

For uniform strings and the mode shapes normalized as in Eq. (3.24), it is shown easily for all i and j that the orthogonality condition and generalized mass, Eqs. (3.41) and (3.42), respectively, reduce to

$$\int_0^\ell \phi_i(x)\phi_j(x)dx = 0 \quad M_i = \frac{m\ell}{2} \quad \text{only for } m = \text{const.} \tag{3.43}$$

3.1.3 Using Orthogonality

The property of orthogonality is useful in many aspects of structural-dynamics analysis. As an illustration, consider the response of an unforced uniform string to initial conditions. In this case, there are no external loads on the string, but it is presumed to have an initial deflection shape and an initial velocity distribution. Let these initial conditions be represented as

$$v(x, 0) = f(x)$$
$$\frac{\partial v}{\partial t}(x, 0) = g(x) \tag{3.44}$$

where both $f(x)$ and $g(x)$ must be compatible with the boundary conditions.

Using Eq. (3.21), these initial conditions can be written in terms of the modal representation as

$$v(x, 0) = \sum_{i=1}^{\infty} F_i \sin \left(\frac{i\pi x}{\ell} \right) = f(x)$$

$$\frac{\partial v}{\partial t}(x, 0) = \sum_{i=1}^{\infty} \frac{E_i i\pi}{\ell} \sqrt{\frac{T}{m}} \sin \left(\frac{i\pi x}{\ell} \right) = g(x)$$

(3.45)

Both of these relationships are multiplied by $\sin(j\pi x/\ell)\,dx$ and integrated over the length of the string. The first relationship yields

$$\int_0^{\ell} f(x) \sin \left(\frac{j\pi x}{\ell} \right) dx = \sum_{i=1}^{\infty} F_i \int_0^{\ell} \sin \left(\frac{i\pi x}{\ell} \right) \sin \left(\frac{j\pi x}{\ell} \right) dx$$

$$= F_j \int_0^{\ell} \sin^2 \left(\frac{j\pi x}{\ell} \right) dx$$

(3.46)

$$= \frac{F_j \ell}{2}$$

where the evaluation used the orthogonality property of the mode shapes, which causes every term in the infinite sum to be zero except where $i = j$. The mass per unit length is constant for this case and hence does not appear under the integral. The second relationship can be reduced in a similar manner, so that

$$\int_0^{\ell} g(x) \sin \left(\frac{j\pi x}{\ell} \right) dx = \sum_{i=1}^{\infty} \frac{E_i i\pi}{\ell} \sqrt{\frac{T}{m}} \int_0^{\ell} \sin \left(\frac{i\pi x}{\ell} \right) \sin \left(\frac{j\pi x}{\ell} \right) dx$$

$$= \frac{E_j j\pi}{\ell} \sqrt{\frac{T}{m}} \int_0^{\ell} \sin^2 \left(\frac{j\pi x}{\ell} \right) dx$$

(3.47)

$$= \frac{E_j j\pi}{2} \sqrt{\frac{T}{m}}$$

This treatment of the initial conditions therefore permits a direct evaluation of the unknown constants (E_i and F_i) in the modal representation of the total string displacement; that is

$$E_i = \frac{2}{i\pi} \sqrt{\frac{m}{T}} \int_0^{\ell} g(x) \sin \left(\frac{i\pi x}{\ell} \right) dx$$

$$F_i = \frac{2}{\ell} \int_0^{\ell} f(x) \sin \left(\frac{i\pi x}{\ell} \right) dx$$

(3.48)

Thus, for the prescribed initial conditions given by $f(x)$ and $g(x)$, the resulting string displacement can be described as

$$v(x, t) = \sum_{i=1}^{\infty} \sin \left(\frac{i\pi x}{\ell} \right) [E_i \sin(\omega_i t) + F_i \cos(\omega_i t)]$$

(3.49)

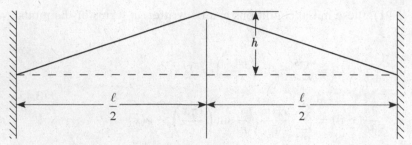

Figure 3.2. Initial shape of plucked string

Example: Response Due to Given Initial Shape. To further illustrate this procedure, consider the case of the plucked string with zero initial velocity. Let the initial shape be as shown in Fig. 3.2. If we assume the initial velocity to be zero, then $g(x) = 0$ and $E_i = 0$ for all i. The string displacement becomes

$$v(x, t) = \sum_{i=1}^{\infty} F_i \sin\left(\frac{i\pi x}{\ell}\right) \cos(\omega_i t) \tag{3.50}$$

To evaluate the constants, F_i, the initial string shape is written as

$$\begin{aligned} f(x) &= 2h\left(\tfrac{x}{\ell}\right) & 0 \le x \le \tfrac{\ell}{2} \\ &= 2h\left(1 - \tfrac{x}{\ell}\right) & \tfrac{\ell}{2} \le x \le \ell \end{aligned} \tag{3.51}$$

Substitution of this function into the preceding integral yields

$$\begin{aligned} F_i &= \frac{4h}{\ell^2}\left[\int_0^{\frac{\ell}{2}} x \sin\left(\frac{i\pi x}{\ell}\right) dx + \int_{\frac{\ell}{2}}^{\ell} (\ell - x) \sin\left(\frac{i\pi x}{\ell}\right) dx\right] \\ &= \frac{8h}{(i\pi)^2} \sin\left(\frac{i\pi}{2}\right) \end{aligned} \tag{3.52}$$

It may be noted that $\sin(i\pi/2)$ is zero for all even values of the index and that it is either $+1$ or -1 for odd values. If desired, these constants can be written as

$$F_i = \begin{cases} \dfrac{8h}{(i\pi)^2}(-1)^{\frac{i-1}{2}} & (i \text{ odd}) \\ 0 & (i \text{ even}) \end{cases} \tag{3.53}$$

The fact that $F_i = 0$ for all even values of i is indicative of the symmetry of the initial string displacement about the midpoint. That is, because the initial shape is symmetric about $x = \ell/2$, no antisymmetric modes of vibration are thereby excited. The total string displacement becomes

$$v(x, t) = \frac{8h}{\pi^2} \sum_{i=1,3,\ldots}^{\infty} \frac{(-1)^{\frac{i-1}{2}}}{i^2} \sin\left(\frac{i\pi x}{\ell}\right) \cos(\omega_i t) \tag{3.54}$$

where

$$\omega_i = \frac{i\pi}{\ell}\sqrt{\frac{T}{m}} \tag{3.55}$$

It should be noted from this solution that the modal contributions to the total displacement significantly decrease as the mode number (i.e., the index i) increases. This can be observed by the dependence of F_i on i and is characteristic of almost all structural-dynamics response problems; thus, it permits a truncation of the infinite sum to a finite number of the lower-order modes. This solution indicates that the string will vibrate forever, with the string periodically returning to its initial shape. In actual systems, however, there are always dissipative phenomena that cause the motion to die out in time. This is considered when we address aeroelastic flutter in Chapter 5.

3.1.4 Traveling Wave Solution

In the preceding section, a modal solution was obtained for the string problem. The solution depicted the total displacement as a summation of specific shapes as measured relative to the ends of the string. Each shape had an amplitude that was, in general, a function of time. When these individual modal contributions were of constant amplitude at their modal frequency, they appeared as standing or fixed waves along the string.

Another interpretation of the string response is now considered by examining the solution obtained for a string with an initial displacement but zero initial velocity and external loading. In this case, the E_is were all zero so that the displacement was written as

$$v(x, t) = \sum_{i=1}^{\infty} \sin\left(\frac{i \pi x}{\ell}\right) F_i \cos\left(\sqrt{\frac{T}{m}} \frac{i \pi t}{\ell}\right) \tag{3.56}$$

The F_is can be determined from the initial shape, $f(x)$, as

$$F_i = \frac{2}{\ell} \int_0^{\ell} f(x) \sin\left(\frac{i \pi x}{\ell}\right) dx \tag{3.57}$$

It also may be noted that the initial shape can be represented by

$$v(x, 0) = f(x) = \sum_{i=1}^{\infty} F_i \sin\left(\frac{i \pi x}{\ell}\right) \tag{3.58}$$

Equation (3.58) is known as the Fourier sine series representation of the function $f(x)$. Additional information on the Fourier series may be found in more advanced textbooks on structural dynamics and applied mathematics. Now, to rewrite the general solution for this problem, the two well-known identities

$$\sin(\alpha + \beta) = \sin(\alpha)\cos(\beta) + \cos(\alpha)\sin(\beta)$$

$$\sin(\alpha - \beta) = \sin(\alpha)\cos(\beta) - \cos(\alpha)\sin(\beta) \tag{3.59}$$

can be added to yield another identity as

$$\sin(\alpha)\cos(\beta) = \frac{1}{2}\left[\sin(\alpha + \beta) + \sin(\alpha - \beta)\right] \tag{3.60}$$

This identity can be used to rewrite the general solution given by Eq. (3.56) as

$$v(x, t) = \frac{1}{2} \sum_{i=1}^{\infty} F_i \left\{ \sin \left[\frac{i\pi}{\ell} \left(x + \sqrt{\frac{T}{m}} t \right) \right] + \sin \left[\frac{i\pi}{\ell} \left(x - \sqrt{\frac{T}{m}} t \right) \right] \right\} \qquad (3.61)$$

Equation (3.58) gives the functional form of $f(x)$ as an infinite sum of sine functions with coefficients F_i. The two terms on the right-hand side of Eq. (3.61) are of the same form as the sum in Eq. (3.58) and can be identified as having the functional form of $f(x)$ but with different arguments. It is therefore possible to rewrite Eq. (3.61) as

$$v(x, t) = \frac{1}{2} \left[f \left(x + \sqrt{\frac{T}{m}} t \right) + f \left(x - \sqrt{\frac{T}{m}} t \right) \right] \qquad (3.62)$$

This is the principal result of the traveling-wave solution. In reality, it is mathematically identical to the previously given standing-wave solution in Eq. (3.56); the only difference is point of view.

To illustrate how Eq. (3.62) represents traveling waves along the string, the two arguments of the shape function are replaced by new spatial coordinates, the origins of which are time dependent. The new coordinates are defined as

$$x_L(x, t) \equiv x + \sqrt{\frac{T}{m}} t$$
$$\qquad (3.63)$$
$$x_R(x, t) \equiv x - \sqrt{\frac{T}{m}} t$$

Equation (3.62) becomes

$$v(x, t) = \frac{1}{2} [f(x_L) + f(x_R)] \qquad (3.64)$$

which indicates that the time-dependent string shape is the sum of two shapes of a form identical to the initial shape but of one half its magnitude. Initially, at $t = 0$, the origins of the x_L and x_R coincide with the x origin as

$$x_L(x, 0) = 0 \quad \text{at} \quad x = 0$$
$$\qquad (3.65)$$
$$x_R(x, 0) = 0 \quad \text{at} \quad x = 0$$

At any later time $t > 0$, the origins of x_L and x_R can be located by

$$x_L(x, t) = 0 \quad \text{at} \quad x = -\sqrt{\frac{T}{m}} t$$
$$\qquad (3.66)$$
$$x_R(x, t) = 0 \quad \text{at} \quad x = \sqrt{\frac{T}{m}} t$$

These results indicate that the x_L coordinate system is moving to the left with a speed $\sqrt{T/m}$ and the x_R coordinate system is moving to the right with the same speed. These origin positions are indicated in Fig. 3.3. As a consequence of these moving origins, the shape $f(x_L)/2$ appears to propagate to the left and the shape

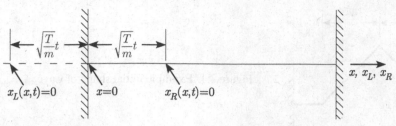

Figure 3.3. Schematic of moving coordinate systems x_L and x_R

$f(x_R)/2$ appears to propagate to the right. Both of these shapes will move at a constant propagation speed of

$$V = \sqrt{\frac{T}{m}} \qquad (3.67)$$

so that Eq. (3.62) may be written in the form

$$v(x, t) = \frac{1}{2}[f(x + Vt) + f(x - Vt)] \qquad (3.68)$$

This is also called D'Alembert's form of the equation.

When these shapes reach one of the walls, the deflection must go to zero to satisfy the boundary conditions. This condition at each wall causes the shapes to be reflected in the opposite direction. These reflections appear as inverted shapes propagating away from the walls, again with the speed $V = \sqrt{T/m}$. This reflected-wave behavior is inherent to the Fourier sine series representation of $f(x)$ given in Eq. (3.58). Determination of the string displacement at times subsequent to $t = 0$ requires the evaluation of $f(x \pm Vt)$ in Eq. (3.62). Although the function $f(x)$ is defined only for the range $0 \leq x \leq \ell$, the arguments $x + Vt$ and $x - Vt$ significantly exceed this range. The Fourier sine series for $f(x)$, Eq. (3.58), possesses two distinct mathematical properties that permit evaluation of the function throughout the extended range of the argument and demonstrate the reflected-wave behavior.

First Property of $f(x)$. Because all terms in the Fourier sine series for $f(x)$ are odd functions of x, $f(x)$ must also be an odd function. This property can be described as

$$f(-x) = -f(x) \qquad (3.69)$$

It is immediately seen that this is a description of the reflected-wave behavior at the $x = 0$ wall.

Second Property of $f(x)$. Because all terms in the Fourier sine series for $f(x)$ are periodic in x with a period of 2ℓ, then $f(x)$ also must be periodic in x with a period of 2ℓ. This property can be described as

$$f(x) = f(x + 2n\ell) \quad \text{for } n = 0, \pm1, \pm2, \ldots \qquad (3.70)$$

This relationship, in conjunction with the previously noted "odd" functionality of $f(x)$, describes the reflected-wave behavior at the $x = \ell$ wall.

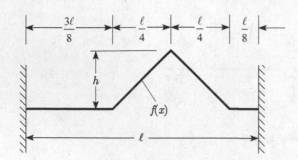

Figure 3.4. Example initial shape of wave

General Evaluation of $f(x \pm Vt)$. These two properties can be applied simultaneously for the evaluation of $f(x + Vt)$ and $f(x - Vt)$ for any value of their argument—say, $x \pm Vt$. When this argument lies within the range

$$n\ell \leq x \pm Vt \leq (n+1)\ell \tag{3.71}$$

where

$$n = 0, \pm 1, \pm 2, \ldots \tag{3.72}$$

then

$$f(x \pm Vt) = (-1)^n f \left\{ (-1)^n \left[x \pm Vt + \frac{(-1)^n - 2n - 1}{2}\ell \right] \right\} \tag{3.73}$$

We used Eq. (3.70) to reduce the range of motion, which was initially $-\infty \leq x \leq +\infty$, down to the range $0 \leq x \leq \ell$, our physical space (i.e., where the string actually is mounted).

Example of Traveling Wave. The initial string shape is given in Fig. 3.4. At subsequent times, the string shape appears as shown in Fig. 3.5. The absolute distance each of the half shapes has traveled at time t is denoted by $\bar{x}$. The faint lines are the displacements associated with the two constituent waves after transformation to bring them into the range $0 \leq x \leq \ell$, and the bold line is the sum of these two displacements. The displacement during the time $\ell\sqrt{m/T} \leq t \leq 2\ell\sqrt{m/T}$ is a mirror image of the progression revealed in Fig. 3.5 with a return to the original shape at $t = 2\ell\sqrt{m/T}$. The motion is periodic thereafter with period $2\ell\sqrt{m/T}$.

3.1.5 Generalized Equations of Motion

Once the free-vibration modes have been determined for a linear, conservative system, it is a straightforward procedure to determine the system's response to any external loading. This is accomplished by treating each mode of vibration as a dimensional degree of freedom whose scalar coordinate is the mode's generalized coordinate. For each of these modal degrees of freedom, a "generalized equation of motion" can be formulated from Lagrange's equations (see the Appendix and Section 2.1.5). The generalized equations of motion for the string problem can be formulated by substituting expressions for the potential and kinetic energies into

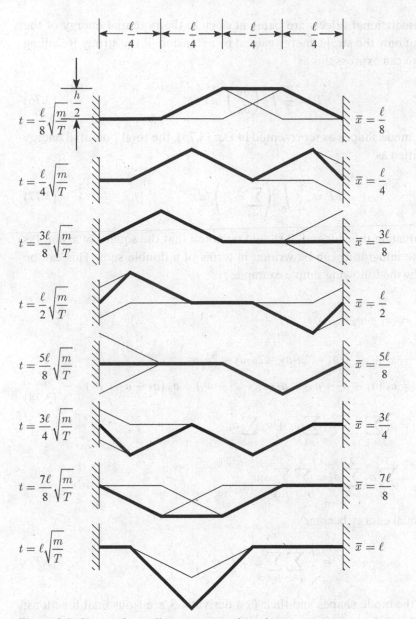

Figure 3.5. Shape of traveling wave at various times

Lagrange's equations; see Eq. (2.14), repeated here for convenience as

$$\frac{d}{dt}\left(\frac{\partial K}{\partial \dot{\xi}_i}\right) + \frac{\partial P}{\partial \xi_i} = \Xi_i \quad (i = 1, 2, \ldots) \tag{3.74}$$

In the energy expressions, the string displacement is represented in terms of its generalized coordinates and mode shapes as

$$v(x, t) = \sum_{i=1}^{\infty} \phi_i(x)\xi_i(t) \tag{3.75}$$

Because gravitational effects are being neglected, the potential energy of the string consists of only the strain energy caused by extension of the string. Recalling Section 2.2.2, we can express this as

$$P = \frac{T}{2} \int_0^\ell \left(\frac{\partial v}{\partial x} \right)^2 dx \tag{3.76}$$

In terms of the mode shapes as represented in Eq. (3.75), the total potential energy then can be written as

$$P = \frac{T}{2} \int_0^\ell \left(\sum_{i=1}^\infty \phi_i' \xi_i \right)^2 dx \tag{3.77}$$

Before evaluating this integral, it should be noted that the square of a sum (as appearing in the integrand) can be written in terms of a double sum. This can be demonstrated by the following simple example:

$$\begin{aligned}
\left(\sum_{i=1}^3 a_i \right)^2 &= (a_1 + a_2 + a_3)^2 \\
&= a_1^2 + a_2^2 + a_3^2 + 2a_1 a_2 + 2a_2 a_3 + 2a_3 a_1 \\
&= a_1 (a_1 + a_2 + a_3) + a_2 (a_1 + a_2 + a_3) + a_3 (a_1 + a_2 + a_3) \\
&= a_1 \sum_{i=1}^3 a_i + a_2 \sum_{i=1}^3 a_i + a_3 \sum_{i=1}^3 a_i \\
&= \sum_{j=1}^3 a_j \sum_{i=1}^3 a_i = \sum_{i=1}^3 \sum_{j=1}^3 a_i a_j
\end{aligned} \tag{3.78}$$

Thus, the potential energy becomes

$$P = \frac{T}{2} \sum_{i=1}^\infty \sum_{j=1}^\infty \xi_i \xi_j \int_0^\ell \phi_i' \phi_j' dx \tag{3.79}$$

For the string, the mode shapes and their first derivatives are sinusoidal functions; consequently, they form an orthogonal set.[2] That is

$$\int_0^\ell \phi_i'(x) \phi_j'(x) dx = 0 \quad (i \neq j) \tag{3.80}$$

Thus, the potential energy relationship can be simplified to

$$P = \frac{T}{2} \sum_{i=1}^\infty \xi_i^2 \int_0^\ell \phi_i'^2 dx \tag{3.81}$$

---

[2] It is *not* true in general that the derivatives of mode-shape functions form an orthogonal set.

The integral in this expression can be integrated by parts as

$$\int_0^\ell \phi_i' \phi_i' dx = \phi_i' \phi_i \Big|_0^\ell - \int_0^\ell \phi_i(x)\phi_i''(x)dx \tag{3.82}$$

By virtue of the boundary conditions at both ends, the first term is zero. Substitution of Eq. (3.34) into the last term (i.e., the integral) shows that

$$T \int_0^\ell \phi_i'^2 dx = \omega_i^2 \int_0^\ell m(x)\phi_i^2 dx = M_i \omega_i^2 \tag{3.83}$$

where we recall that ω_i is the natural frequency of the ith mode and that M_i is the generalized mass (see Eq. [3.42]). (Note that the ith generalized mass depends on the mode shape of the ith mode and on how that mode shape is normalized.) Thus, the potential energy becomes

$$P = \frac{1}{2} \sum_{i=1}^\infty M_i \omega_i^2 \xi_i^2 \tag{3.84}$$

Recalling the kinetic energy from Eq. (2.37), repeated here for convenience as

$$K = \int_0^\ell \left[\frac{m}{2} \left(\frac{\partial v}{\partial t} \right)^2 \right] dx \tag{3.85}$$

we may now use the modal representation to write

$$K = \frac{1}{2} \int_0^\ell m \left(\sum_{i=1}^\infty \phi_i \dot{\xi}_i \right)^2 dx \tag{3.86}$$

With the double-sum notation, the kinetic energy simplifies to

$$K = \frac{1}{2} \int_0^\ell \sum_{i=1}^\infty \sum_{j=1}^\infty \phi_i \dot{\xi}_i \phi_j \dot{\xi}_j m(x) \, dx$$

$$= \frac{1}{2} \sum_{i=1}^\infty \sum_{j=1}^\infty \dot{\xi}_i \dot{\xi}_j \int_0^\ell m(x)\phi_i \phi_j dx \tag{3.87}$$

Because the mode shapes are orthogonal functions where

$$\int_0^\ell m(x)\phi_i(x)\phi_j(x)dx = \begin{cases} 0 & (i \neq j) \\ M_i \neq 0 & (i = j) \end{cases} \tag{3.88}$$

the total kinetic energy becomes

$$K = \frac{1}{2} \sum_{i=1}^\infty M_i \dot{\xi}_i^2 \tag{3.89}$$

The "generalized equations of motion" now can be obtained by substitution of the kinetic energy of Eq. (3.89) and the potential energy of Eq. (3.84) into Lagrange's equations given as Eqs. (3.74). The resulting equations are then

$$M_i \left(\ddot{\xi}_i + \omega_i^2 \xi_i \right) = \Xi_i \quad (i = 1, 2, \ldots) \tag{3.90}$$

When using a modal representation, we may use equations of this form for the dynamic analysis of any linearly elastic structure. The generalized mass and natural frequencies, of course, will differ depending on whether the structure is a string, a beam in torsion or bending, a plate or shell, or a complete aircraft. The left-hand side of this equation has at least these terms regardless of the system being analyzed. Kinetic and potential energies also may contain contributions from discrete elements such as added particles, rigid bodies, or springs. Finally, additional terms could arise from potential energy of conservative applied loads, such as gravity.

The right-hand side, conversely, is highly problem-dependent and is addressed next. The case of a vibrating structure without external forces is a special case, previously discussed in Section 3.1.3. When there are no external forces, $\Xi_i = 0$ for all i. The resulting general solution of Eq. (3.90) is the same as that presented in Section 3.1.3, which was obtained without reference to the generalized equations of motion and yields results depending only on the initial conditions. When including an entity such as a spring in the potential energy, we are enlarging the boundary of the system to include a new element. However, when the same entity is included through its contribution to the generalized forces, it is being treated as a source of external forces, something external to the system. Despite this philosophical distinction, the end result is the same (see Problem 5). Any effect that can be included in the generalized equations of motion through potential energy can be included instead through the generalized force. It is extremely important to not count the same effect twice (e.g., including the same entity through both potential energy and generalized forces).

3.1.6 Generalized Force

The generalized force, $\Xi_i(t)$—which appears on the right-hand side of the generalized equations of motion—represents the effective loading associated with all forces and moments not accounted for in P, which includes any nonconservative forces and moments. These forces and moments are most commonly identified as externally applied loads, which may or may not be a function of modal response. They also include any dissipative loads such as those from dampers. To determine the contribution of distributed loads, denoted by $f(x, t)$, the virtual work is computed from Eq. (2.38), repeated here for convenience as

$$\overline{\delta W} = \int_0^\ell f(x, t) \delta v(x, t) dx \tag{3.91}$$

The term $\delta v(x, t)$ represents a variation of the displacement field, typically referred to as the "virtual displacement," which can be written in terms of the

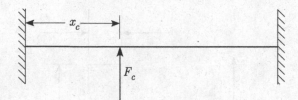

Figure 3.6. Concentrated force acting on string

generalized coordinates and mode shapes as

$$\delta v(x, t) = \sum_{i=1}^{\infty} \phi_i(x)\delta\xi_i(t) \tag{3.92}$$

where $\delta\xi_i(t)$ is an arbitrary increment in the ith generalized coordinate. Thus, the virtual work becomes

$$\overline{\delta W} = \int_0^\ell \sum_{i=1}^{\infty} f(x, t)\phi_i(x)\delta\xi_i(t)dx$$
$$= \sum_{i=1}^{\infty} \delta\xi_i(t) \int_0^\ell f(x, t)\phi_i(x)dx \tag{3.93}$$

Identifying the generalized force as

$$\Xi_i(t) = \int_0^\ell f(x, t)\phi_i(x)dx \tag{3.94}$$

we find that the virtual work reduces to

$$\overline{\delta W} = \sum_{i=1}^{\infty} \Xi_i(t)\,\delta\xi_i(t) \tag{3.95}$$

The loading $f(x, t)$ in this development is a distributed load with units of force per unit length. If instead this loading is concentrated at one or more points—say, as $F_c(t)$ with units of force acting at $x = x_c$ as shown in Fig. 3.6—then its functional representation must include the Dirac delta function, $\delta(x - x_c)$, which is similar to the impulse function in the time domain. In this case, the distributed load can be written as

$$f(x, t) = F_c(t)\delta(x - x_c) \tag{3.96}$$

Recall that the Dirac delta function can be thought of as the limiting case of a rectangular shape with area held constant and equal to unity as its width goes to zero (Fig. 3.7). Thus, it may be defined by its integral property; for example, for $a < x_0 < b$

$$\int_a^b \delta(x - x_0)dx = 1$$
$$\int_a^b f(x)\delta(x - x_0)dx = f(x_0) \tag{3.97}$$

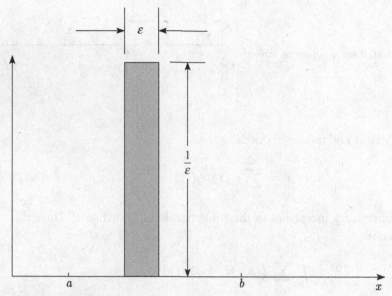

Figure 3.7. Approaching the Dirac delta function

As a consequence, this integral expression for the generalized force can be applied to the concentrated load so that

$$
\begin{aligned}
\Xi_i(t) &= \int_0^\ell F_c(t)\delta(x - x_c)\phi_i(x)dx \\
&= F_c(t) \int_0^\ell \delta(x - x_c)\phi_i(x)dx \\
&= F_c(t)\phi_i(x_c) \\
&= F_c(t)\sin\left(\frac{i\pi x_c}{\ell}\right)
\end{aligned}
\tag{3.98}
$$

3.1.7 Example Calculations of Forced Response

In this section, we present two examples of forced-response calculations. These examples also appropriately are called "initial-value problems." The first has zero initial displacement and velocity; the second has nonzero initial displacement and zero initial velocity.

Example: Calculation of Forced Response. An example of a dynamically loaded uniform string is considered to illustrate the generalized force computation and subsequent solution for the string displacement. The specific example is a uniformly distributed load (in space) of simple harmonic amplitude (in time) shown in Fig. 3.8 with

$$
f(x, t) = \overline{F}\sin(\omega t)
\tag{3.99}
$$

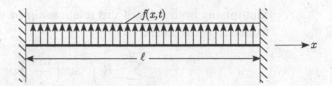

Figure 3.8. Distributed force $f(x, t)$ acting on string

The initial string displacement and velocity are taken as zero. Computation of the generalized force is simply

$$\Xi_i = \int_0^\ell \overline{F} \sin(\omega t) \sin\left(\frac{i\pi x}{\ell}\right) dx$$

$$= \frac{\overline{F}\ell}{i\pi} \sin(\omega t)[1 - \cos(i\pi)]$$

(3.100)

Considering the even- and odd-indexed modes separately, we have

$$\Xi_i = \begin{cases} \dfrac{2\overline{F}\ell}{i\pi} \sin(\omega t) & (i \text{ odd}) \\ 0 & (i \text{ even}) \end{cases}$$

(3.101)

With this equation, the generalized equations of motion become

$$M_i(\ddot{\xi}_i + \omega_i^2 \xi_i) = \begin{cases} \dfrac{2\overline{F}\ell}{i\pi} \sin(\omega t) & (i \text{ odd}) \\ 0 & (i \text{ even}) \end{cases}$$

(3.102)

Because the initial conditions on displacement and velocity are both identically zero, that is

$$v(x, 0) = \frac{\partial v}{\partial t}(x, 0) = 0$$

(3.103)

it follows that the response is governed only by the generalized forces. Thus, the even-indexed modes are not excited because their generalized forces are also zero. For the odd-indexed modes, the general solution to their equation of motion is

$$\xi_i = A_i \sin(\omega_i t) + B_i \cos(\omega_i t) + C_i \sin(\omega t)$$

(3.104)

Note that the first two terms correspond to the homogeneous portion of the solution, whereas the third term represents the particular solution. In this example, the particular solution has the same form of time dependence as the generalized force.

To evaluate the constants A_i and B_i of the homogeneous solution, a procedure can be followed that is similar to the one used in Section 3.1.2 for solution of the homogeneous initial-condition problem. The initial displacement of the present example can be written as

$$v(x, 0) = \sum_{i=1,3,\dots}^\infty \phi_i(x)\xi_i(0) = \sum_{i=1,3,\dots}^\infty B_i \sin\left(\frac{i\pi x}{\ell}\right) = 0$$

(3.105)

Multiplying both sides of this relationship by $\sin(j\pi x/\ell)dx$ and integrating over x from 0 to ℓ, we obtain

$$\sum_{i=1,3,\ldots}^{\infty} B_i \int_0^{\ell} \sin\left(\frac{i\pi x}{\ell}\right) \sin\left(\frac{j\pi x}{\ell}\right) dx = 0 \qquad (3.106)$$

Applying the orthogonality property of the sine functions in the integrand indicates that

$$B_i = 0 \quad (i \text{ odd}) \qquad (3.107)$$

The same procedure can be applied to the initial velocity, where

$$\frac{\partial v}{\partial t}(x,0) = \sum_{i=1,3,\ldots}^{\infty} \phi_i(x)\dot{\xi}_i(0) = \sum_{i=1,3,\ldots}^{\infty} (A_i\omega_i + C_i\omega) \sin\left(\frac{i\pi x}{\ell}\right) = 0 \qquad (3.108)$$

Again, this relationship can be multiplied by $\sin(j\pi x/\ell)dx$ and integrated over the string length. The orthogonality property in this case yields

$$A_i = -\frac{\omega C_i}{\omega_i} \quad (i \text{ odd}) \qquad (3.109)$$

Initial conditions of zero displacement and velocity thus require that the generalized coordinates of the odd-indexed modes be written as

$$\xi_i = C_i \left[\sin(\omega t) - \frac{\omega}{\omega_i} \sin(\omega_i t) \right] \quad (i \text{ odd}) \qquad (3.110)$$

The constants C_i of the particular solution can be determined by substitution of the generalized coordinate back into the generalized equations of motion. This yields

$$M_i C_i \left(\omega_i^2 - \omega^2 \right) \sin(\omega t) = \frac{2\overline{F}\ell}{i\pi} \sin(\omega t) \qquad (3.111)$$

Using Eq. (3.42), we find that $M_i = m\ell/2$ for all i. Thus, C_i becomes

$$C_i = \frac{4\overline{F}}{i\pi m \left(\omega_i^2 - \omega^2 \right)} \qquad (3.112)$$

Thus, the string displacement can be now written as the sum of contributions from the odd-indexed modes. Recall that neither the excitation loading nor the initial conditions excite the even-indexed modes. Thus

$$\begin{aligned} v(x,t) &= \sum_{i=1,3,\ldots}^{\infty} \xi_i(t)\phi_i(x) \\ &= \frac{4\overline{F}}{m\pi} \sum_{i=1,3,\ldots}^{\infty} \left[\frac{\sin(\omega t) - \frac{\omega}{\omega_i} \sin(\omega_i t)}{i \left(\omega_i^2 - \omega^2 \right)} \right] \sin\left(\frac{i\pi x}{\ell}\right) \end{aligned} \qquad (3.113)$$

When the forcing frequency coincides with one of the natural frequencies, an interesting situation results. Considering only the time-dependent part of a typical

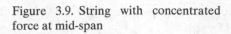

Figure 3.9. String with concentrated force at mid-span

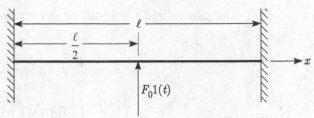

term in the series of Eq. (3.113), that is

$$\frac{\sin(\omega t) - \frac{\omega}{\omega_i} \sin(\omega_i t)}{i \left(\omega_i^2 - \omega^2\right)} \tag{3.114}$$

we find that when $\omega \to \omega_i$, the term becomes indeterminate. To see what its value is in the limit, we let $\omega_i = \omega + \varepsilon_i$, which gives

$$\frac{\sin(\omega t) - \frac{\omega}{\omega+\varepsilon_i} \sin\left[(\omega + \varepsilon_i)\, t\right]}{i \left[(\omega + \varepsilon_i)^2 - \omega^2\right]} \tag{3.115}$$

Invoking l'Hopital's rule to take the limit as $\varepsilon_i \to 0$, we obtain

$$\frac{\sin(\omega t) - \omega t \cos(\omega t)}{2i\omega^2} \tag{3.116}$$

The second term tends to infinity as time increases with a linearly increasing amplitude. This phenomenon is called "resonance" and, because of its destructive nature, should be avoided. That is, when a structure is excited using harmonic excitation, the forcing frequency must not be too near any of the structure's natural frequencies.

Example: Calculation of Forced Response with Nonzero-Initial Conditions. A second example is considered to illustrate the treatment of a concentrated force and initial conditions that are not identically zero. In this case, a concentrated step-function force of magnitude F_0 is applied to the center of the string, as illustrated in Fig. 3.9. Recall that the unit-step function, $1(t)$, is defined by

$$
\begin{aligned}
1(t) &= 0 \quad (t < 0) \\
&= 1 \quad (t \geq 0)
\end{aligned} \tag{3.117}
$$

The initial shape of the string is given as

$$v(x, 0) = h \sin\left(\frac{4\pi x}{\ell}\right) \tag{3.118}$$

and the initial velocity as zero.

The generalized force can be determined from the integral of a distributed loading as

$$
\begin{aligned}
\Xi_i &= \int_0^\ell f(x,t)\phi_i(x)dx \\
&= \int_0^\ell F_0 1(t)\delta\left(x - \frac{\ell}{2}\right)\phi_i(x)dx \\
&= F_0 1(t)\phi_i\left(\frac{\ell}{2}\right) \\
&= F_0 1(t)\sin\left(\frac{i\pi}{2}\right)
\end{aligned}
\tag{3.119}
$$

because

$$
\begin{aligned}
\sin\left(\frac{i\pi}{2}\right) &= 0 && (i\ \text{even}) \\
&= (-1)^{\frac{i-1}{2}} && (i\ \text{odd})
\end{aligned}
\tag{3.120}
$$

the generalized equations of motion become

$$
\begin{aligned}
M_i(\ddot{\xi}_i + \omega_i^2 \xi_i) &= 0 && (i\ \text{even}) \\
M_i(\ddot{\xi}_i + \omega_i^2 \xi_i) &= F_0 1(t)(-1)^{\frac{i-1}{2}} && (i\ \text{odd})
\end{aligned}
\tag{3.121}
$$

The corresponding general solutions are

$$
\begin{aligned}
\xi_i &= A_i \sin(\omega_i t) + B_i \cos(\omega_i t) && (i\ \text{even}) \\
\xi_i &= A_i \sin(\omega_i t) + B_i \cos(\omega_i t) + C_i && (i\ \text{odd})
\end{aligned}
\tag{3.122}
$$

Consider the finite, initial displacement

$$
\begin{aligned}
v(x,0) &= \sum_{i=1}^\infty \xi_i(0)\phi_i(x) \\
&= \sum_{i=2,4,\ldots}^\infty B_i \sin\left(\frac{i\pi x}{\ell}\right) + \sum_{i=1,3,\ldots}^\infty (B_i + C_i)\sin\left(\frac{i\pi x}{\ell}\right) \\
&= h\sin\left(\frac{4\pi x}{\ell}\right)
\end{aligned}
\tag{3.123}
$$

This last equality is multiplied by $\sin(j\pi x/\ell)dx$ and integrated over the length of the string to yield

$$
\begin{aligned}
h\int_0^\ell \sin\left(\frac{4\pi x}{\ell}\right)\sin\left(\frac{j\pi x}{\ell}\right)dx &= \sum_{i=2,4,\ldots}^\infty B_i \int_0^\ell \sin\left(\frac{i\pi x}{\ell}\right)\sin\left(\frac{j\pi x}{\ell}\right)dx \\
&\quad + \sum_{i=1,3,\ldots}^\infty (B_i + C_i)\int_0^\ell \sin\left(\frac{i\pi x}{\ell}\right)\sin\left(\frac{j\pi x}{\ell}\right)dx
\end{aligned}
\tag{3.124}
$$

These integrals can be evaluated easily by noting the orthogonality property of the sine functions. The result gives the following values for the constants B_i:

$$B_4 = h$$

$$B_i = 0 \quad (i \text{ even but } \neq 4) \tag{3.125}$$

$$B_i = -C_i \quad (i \text{ odd})$$

The initial velocity being zero requires that

$$\frac{\partial v}{\partial t}(x, 0) = \sum_{i=1}^{\infty} \dot{\xi}_i(0)\phi_i(x) = \sum_{i=1}^{\infty} \omega_i A_i \sin\left(\frac{i\pi x}{\ell}\right) = 0 \tag{3.126}$$

Multiplication by $\sin(j\pi x/\ell)dx$ and integration results in determining that $A_i = 0$ for all i. These results can be summarized by noting that $\xi_i = 0$ for all even values of i except

$$\xi_4 = h \cos(\omega_4 t) \tag{3.127}$$

and for odd i

$$\xi_i = C_i [1 - \cos(\omega_i t)] \quad (i \text{ odd}) \tag{3.128}$$

The constants C_i can be determined by substitution of the odd generalized coordinates back into the equations of motion

$$M_i C_i \omega_i^2 = F_0(-1)^{\frac{i-1}{2}} \quad t \geq 0 \tag{3.129}$$

Given that $M_i = m\ell/2$, this yields

$$C_i = \frac{2\ell F_0(-1)^{\frac{i-1}{2}}}{T(i\pi)^2} \tag{3.130}$$

so that the complete string displacement becomes

$$v(x, t) = \sum_{i=1}^{\infty} \xi_i(t)\phi_i(x)$$

$$= h \cos(\omega_4 t) \sin\left(\frac{4\pi x}{\ell}\right) + \frac{2\ell F_0}{T\pi^2} \sum_{i=1,3,\ldots}^{\infty} \frac{(-1)^{\frac{i-1}{2}}}{i^2} [1 - \cos(\omega_i t)] \sin\left(\frac{i\pi x}{\ell}\right) \tag{3.131}$$

Thus, the first term is the response due to initial displacement, and the sum over the odd-indexed modes is the response due to the forcing function.

3.2 Uniform Beam Torsional Dynamics

Although vibrating strings are easy to visualize and exhibit many of the features of vibrating aerospace structures, to analyze such structures, more realistic models are needed. In this section, we apply the concepts related to the modal representation to the dynamics of beams in torsion. A beam is a structural member in which one dimension is much larger than the other two. It is thus understandable to idealize the twisting and bending of high-aspect-ratio wings and helicopter rotor blades in terms of beam theory, especially in conceptual and preliminary design. Because many behavioral characteristics of typical aeronautical structures are found in beams, the torsion of beam-like lifting surfaces plays a vital role in both static and dynamic aeroelasticity.

3.2.1 Equations of Motion

For free vibration of a beam in torsion, we specialize the equation of motion derived in Section 2.3.1 by setting $r(x, t) = 0$ to obtain

$$\frac{\partial}{\partial x}\left[\overline{GJ}(x)\frac{\partial \theta}{\partial x}\right] = \overline{\rho I_p}(x)\frac{\partial^2 \theta}{\partial t^2} \tag{3.132}$$

Other than the quantities that multiply the partial derivatives, this equation of motion is similar to that for the dynamic behavior of a string. The difference is that the stiffness coefficient $\overline{GJ}(x)$, unlike the tension in the string, may not be constant. Specialization to the spanwise uniform case is undertaken to obtain a closed-form solution. Properties varying with x are not an obstacle for application of the variety of approximate methods discussed in Section 3.5, but here we are concerned with obtaining closed-form solutions to aid in understanding the results. As shown when we explore the boundary conditions in detail, there are more interesting possibilities for the boundary conditions for beams in torsion than there are for the string.

As before, we apply separation of variables, by substituting

$$\theta(x, t) = X(x)Y(t) \tag{3.133}$$

into the partial differential equation of motion and arranging the terms so that dependencies on x and t are separated across the equality. This yields

$$\frac{\left[\overline{GJ}(x)X'(x)\right]'}{\overline{\rho I_p}(x)X(x)} = \frac{\ddot{Y}(t)}{Y(t)} \tag{3.134}$$

Thus, each side must equal a constant—say, $-\omega^2$—so that

$$\frac{\left[\overline{GJ}(x)X'(x)\right]'}{\overline{\rho I_p}(x)X(x)} = \frac{\ddot{Y}(t)}{Y(t)} = -\omega^2 \tag{3.135}$$

Two ordinary differential equations then follow; namely

$$\begin{aligned}
&\left[\overline{GJ}(x)X'(x)\right]' + \overline{\rho I_p}(x)\omega^2 X(x) = 0 \\
&\ddot{Y}(t) + \omega^2 Y(t) = 0
\end{aligned} \tag{3.136}$$

The first of Eqs. (3.136) has variable coefficients in x and—except for certain special cases such as spanwise uniformity—does not possess a closed-form solution. The second, however, is the same as the second of Eqs. (3.6), the solution of which is well known.

Some specialization is necessary in order to proceed further. Therefore, we consider only beams with spanwise uniform properties. Eqs. (3.136) then become

$$X'' + \alpha^2 X = 0$$
$$\ddot{Y} + \omega^2 Y = 0 \tag{3.137}$$

where $\alpha^2 = \overline{\rho I_p}\omega^2/\overline{GJ}$. For $\alpha \neq 0$, the solutions can be written as

$$X(x) = A\sin(\alpha x) + B\cos(\alpha x)$$
$$Y(t) = C\sin(\omega t) + D\cos(\omega t) \tag{3.138}$$

To complete the solutions, constants A and B can be determined to within a multiplicative constant from the boundary conditions at the ends of the beam; C and D can be found as a function of the initial beam deflection and rate of deflection. Because the partial differential equations of motion governing both transverse vibration of uniform strings and torsional vibration of uniform beams are one-dimensional wave equations, we rightfully can expect all of the previously discussed properties of standing and traveling waves to exist here as well.

Note that the special case of $\alpha = 0$ is an important special case with a different set of general solutions. It is addressed in more detail in Section 3.2.3.

3.2.2 Boundary Conditions

For a beam undergoing pure torsion, one boundary condition is required at each end. Mathematically, boundary conditions may affect θ as well as its partial derivatives, such as $\partial\theta/\partial x$ and $\partial^2\theta/\partial t^2$, at the ends of the beam. In the context of the separation of variables, these conditions lead to corresponding conditions on X and/or X' at the ends. These relationships are necessary and sufficient for determination of the constants A and B to within a multiplicative constant.

The nature of the boundary condition at an end stems from how that end is restrained. When an end cross section is unrestrained, the tractions on it are identically zero. Conversely, the most stringent condition is a perfect clamp, which allows no rotation of an end cross section. Although this is a common idealization, it is practically impossible to achieve in practice.

Cases that only partially restrain an end cross section involve elastic and/or inertial reactions. For example, an aircraft wing attached to a flexible support, such as a fuselage, is not a perfect clamped condition; the root of the wing experiences some rotation because of inherent flexibility at the point of attachment. A boundary condition that is idealized in terms of a rotational spring may be used to create a more realistic model for the support flexibility. Appropriate values for support flexibility can be estimated from static tests. Boundary conditions involving inertial reactions may stem from attached rigid bodies to model the effects of fuel tanks, engines, armaments, and so on.

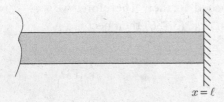

Figure 3.10. Clamped end of a beam

$x = \ell$

In this section, we consider two boundary conditions of the "primitive" type and two examples of derived boundary conditions that can be imposed at the ends of the beam to determine the constants A and B.

Clamped End. In this first primitive case (Fig. 3.10), the $x = \ell$ end of the beam is assumed to be clamped or rigidly attached to an immovable support. As a consequence, there is no rotation due to elastic twist at this end of the beam, and the boundary condition is

$$\theta(\ell, t) = 0 = X(\ell)Y(t) \tag{3.139}$$

which is identically satisfied when

$$X(\ell) = 0 \tag{3.140}$$

Free End. For the second primitive case, we consider the $x = \ell$ end cross section of the beam to be free of stress (Fig. 3.11). Therefore, the twisting moment resultant on the end cross section must be zero

$$T(\ell, t) = \overline{GJ}(\ell)\frac{\partial \theta}{\partial x}(\ell, t) = 0 \tag{3.141}$$

Because $\overline{GJ}(\ell) > 0$, this specializes to

$$\frac{\partial \theta}{\partial x}(\ell, t) = X'(\ell)Y(t) = 0 \tag{3.142}$$

Thus, the specific condition to be satisfied is

$$X'(\ell) = 0 \tag{3.143}$$

Other Forms of End Restraint. In any cross section of a beam undergoing deformation, there is a set of tractions on the plane of a typical cross section; a traction is a projection of the stress (three-dimensional) onto a surface (two-dimensional). From tractions at a given cross section, we can define the resultant force and moment at that station of the beam. When the end of a beam is connected to a rigid body, the

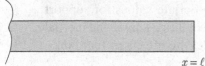

Figure 3.11. Free end of a beam

$x = \ell$

Figure 3.12. Schematic of the $x = \ell$ end of the beam, showing the twisting moment T, and the equal and opposite torque acting on the rigid body

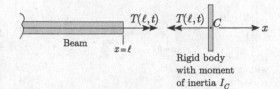

body exerts forces and moments on the beam that are balanced by the distributed traction on the end cross section. That is, the force and moment resultants on the end cross section are reacted by equal and opposite forces and moments on the rigid body. These facts, along with application of suitable laws of motion for the attached body, allow us to determine the boundary conditions.

In the case of torsion, for any cross section, the resultant moment about x of tractions caused by transverse shearing stress has the sense of a twisting moment, T, given by

$$T \equiv \overline{GJ}\frac{\partial \theta}{\partial x} \tag{3.144}$$

With x directed to the right and the outward-directed normal along the positive x direction at the end of the beam where $x = \ell$, a positive twisting moment is directed along x in the right-handed sense. To avoid coupling with transverse motion, we stipulate that the mass center C of the attached rigid body lies on the x axis (i.e., the elastic axis of the beam). The body has a mass moment of inertia I_C about C so that it contributes a concentrated rotational inertia effect on the beam. The free-body diagram for the problem is then as shown in Fig. 3.12. By Newton's third law, the beam's twisting moment produces an equal and opposite torque on the rigid body.

Recall that Euler's second law for a rigid body is stated precisely in Section 2.1.2. The only forces acting on the rigid body here[3] are the contact forces from the beam. So, the x component of the left-hand side of Euler's law is the sum of all moments acting on the body; that is

$$\left(\sum \mathbf{M}_c\right)_x \equiv -T(\ell, t) \tag{3.145}$$

where the sign in front of T is negative because of the free-body diagram and the sign convention, which has moments acting on the body as positive in the same direction as the rotation of the body—along x in the right-handed sense. The x component of the right-hand side is the inertial time derivative of the inertial angular momentum about C, here simply the moment of inertia times the angular acceleration:

$$\left(\frac{^F d\mathbf{H}_c}{dt}\right)_x \equiv I_C \frac{\partial^2 \theta}{\partial t^2}(\ell, t) \tag{3.146}$$

where the left superscript on the left-hand side reflects the fact that the time derivative is taken in the inertial frame F, defined in Section 2.1.2. Euler's law for this rigid

---

[3] Recall that we normally ignore gravity for free-vibration problems.

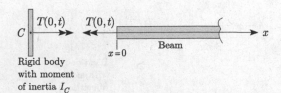

Figure 3.13. Schematic of the $x = 0$ end of the beam, showing the twisting moment T, and the equal and opposite torque acting on the rigid body

body is then expressed by equating these two very different quantities, yielding

$$-T(\ell, t) = I_C \frac{\partial^2 \theta}{\partial t^2}(\ell, t) \tag{3.147}$$

or

$$-\overline{GJ} \frac{\partial \theta}{\partial x}(\ell, t) = I_C \frac{\partial^2 \theta}{\partial t^2}(\ell, t) \tag{3.148}$$

This equation then expresses the boundary condition of a beam undergoing uncoupled torsional vibration with a rigid body attached at $x = \ell$.

When the body is attached to the $x = 0$ end, there is a subtle but important difference. With x directed to the right and the outward-directed normal along the negative x direction at the $x = 0$ end of the beam, a positive twisting moment is directed along $-x$ in the right-handed sense. The free-body diagram for the problem is then as shown in Fig. 3.13. By Newton's third law, the twisting moment produces an equal and opposite torque on the rigid body, which is in the direction of a positive rotation for the body. Therefore, Euler's law (and the resulting boundary condition) is written as

$$T(0, t) = I_C \frac{\partial^2 \theta}{\partial t^2}(0, t) \tag{3.149}$$

or

$$\overline{GJ} \frac{\partial \theta}{\partial x}(0, t) = I_C \frac{\partial^2 \theta}{\partial t^2}(0, t) \tag{3.150}$$

Thus, this equation expresses the boundary condition of a beam undergoing uncoupled torsional vibration with a rigid body attached at $x = 0$.

The following example illustrates a convenient way to think about the contribution of a spring to the boundary of a beam undergoing only torsional rotation. Consider a beam with an attached rigid body at its $x = \ell$ end, such that the rigid body is, in turn, restrained by a light torsional spring attached to the ground. The rotational sign convention does not change; θ is always positive in the x direction (i.e., to the right in the sense of the right-hand rule). The rigid body, because it is attached to the end of the beam, rotates by $\theta(\ell, t)$. Thus, the rotational spring reacts against that rotation, and the moment applied by the spring to the body is opposite to the direction of the rotation (Fig. 3.14). The boundary condition for the beam results from applying Euler's second law to the rigid body, so that

$$-k\theta(\ell, t) - T(\ell, t) = I_C \frac{\partial^2 \theta}{\partial t^2}(\ell, t) \tag{3.151}$$

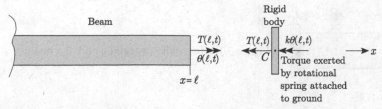

Figure 3.14. Example with rigid body and spring

If the body-spring mechanism is instead on the $x = 0$ end of the beam, then the moment exerted by the spring is still in the same direction. However, what constitutes a positive twisting moment on the beam has the opposite sense, and the boundary condition changes to

$$-k\theta(0, t) + T(0, t) = I_C \frac{\partial^2 \theta}{\partial t^2}(0, t) \tag{3.152}$$

Clearly, if the body is absent, one may set $I_C = 0$. Then, the problem reduces to the elastically restrained boundary condition, as shown in Fig. 3.15. The twisting moment at the beam end must be equal and opposite to the spring reaction for any finite rotation due to twist at the $x = \ell$ end, so that

$$-T(\ell, t) = -\overline{GJ} \frac{\partial \theta}{\partial x}(\ell, t) = k\theta(\ell, t) \tag{3.153}$$

At the $x = 0$ end, however

$$T(0, t) = \overline{GJ} \frac{\partial \theta}{\partial x}(0, t) = k\theta(0, t) \tag{3.154}$$

To be useful for separation of variables, we must determine the corresponding boundary condition on X. Thus, we write $\theta(x, t)$ as $X(x)Y(t)$ as before, yielding

$$\overline{GJ} X'(\ell)Y(t) = -kX(\ell)Y(t) \tag{3.155}$$

which requires that

$$\overline{GJ} X'(\ell) = -kX(\ell) \tag{3.156}$$

Readers should verify that the same type of boundary condition at the other end would yield

$$\overline{GJ} X'(0) = kX(0) \tag{3.157}$$

where the sign change comes about by virtue of the switch in the direction noted previously for a positive twisting moment.

Figure 3.15. Elastically restrained end of a beam

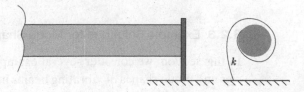

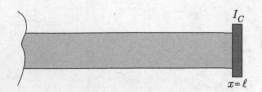

Figure 3.16. Inertially restrained end of a beam

$x = \ell$

Conversely, if the spring is absent, we may set $k = 0$, and the problem reduces to the inertially restrained case with only a rigid body attached to the $x = \ell$ end (Fig. 3.16). The twisting moment at the beam end must be equal and opposite to the inertial reaction of the concentrated inertia for any finite angular acceleration of the end. Therefore

$$-\overline{GJ}\frac{\partial \theta}{\partial x}(\ell, t) = I_C\frac{\partial^2 \theta}{\partial t^2}(\ell, t) \tag{3.158}$$

Expressing this condition in terms of $X(x)$, $Y(t)$, and their derivatives, we find that

$$-\overline{GJ}\,X'(\ell)Y(t) = I_C X(\ell)\ddot{Y}(t) \tag{3.159}$$

From separation of variables, it was determined from the second of Eqs. (3.137) that for free vibration (i.e., no external forces), we may regard $Y(t)$ as describing simple harmonic motion; that is

$$\ddot{Y}(t) = -\omega^2 Y(t) = -\frac{\alpha^2 \overline{GJ}}{\overline{\rho I_p}}Y(t) \tag{3.160}$$

Substitution into the preceding condition then yields

$$\overline{GJ}\,X'(\ell)Y(t) = \alpha^2 \frac{\overline{GJ}}{\overline{\rho I_p}}I_C X(\ell)Y(t) \tag{3.161}$$

which requires that

$$\overline{\rho I_p}\,X'(\ell) = \alpha^2 I_C X(\ell) \tag{3.162}$$

As before, readers should verify that the same type of boundary condition at the other end would yield

$$\overline{\rho I_p}\,X'(0) = -\alpha^2 I_C X(0) \tag{3.163}$$

It is appropriate to note that the use of Eq. (3.160) allows us to express Eq. (3.158) as

$$\overline{GJ}\frac{\partial \theta}{\partial x}(\ell, t) = \omega^2 I_C \theta(\ell, t) \tag{3.164}$$

with the caveat that this condition holds true only for simple harmonic motion.

3.2.3 Example Solutions for Mode Shapes and Frequencies

In this section, we consider several examples of the calculation of natural frequencies and mode shapes of vibrating beams in torsion. We begin with the clamped-free

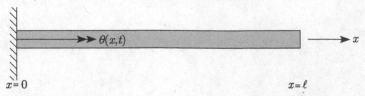

Figure 3.17. Schematic of clamped-free beam undergoing torsion

case, often referred to as "cantilevered." Next, we consider the free-free case, illustrating the concept of the rigid-body mode. Finally, we consider a case that requires numerical solution of the transcendental characteristic equation: a beam clamped at its root and restrained with a rotational spring at the tip.

Example Solution for Clamped-Free Beam. To illustrate the application of these boundary conditions, consider the case of a uniform beam that is clamped at $x = 0$ and free at $x = \ell$, as shown in Fig. 3.17. The boundary conditions for this case are

$$X(0) = X'(\ell) = 0 \tag{3.165}$$

Recall that the general solution was previously determined as

$$\theta(x, t) = X(x)Y(t) \tag{3.166}$$

where X and Y are given in Eqs. (3.138). For $\alpha \neq 0$, the first of those equations has the solution

$$X(x) = A\sin(\alpha x) + B\cos(\alpha x) \tag{3.167}$$

It is apparent that the boundary conditions lead to the following

$$X(0) = 0 \text{ requires } B = 0$$
$$X'(\ell) = 0 \text{ requires } A\alpha \cos(\alpha \ell) = 0 \tag{3.168}$$

If $A = 0$, a trivial solution is obtained, such that the deflection is identically zero. Because $\alpha \neq 0$, a nontrivial solution requires that

$$\cos(\alpha \ell) = 0 \tag{3.169}$$

This is called the "characteristic equation," the solutions of which consist of a denumerably infinite set called the "eigenvalues" and are given by

$$\alpha_i \ell = \frac{(2i - 1)\pi}{2} \quad (i = 1, 2, \ldots) \tag{3.170}$$

The $Y(t)$ portion of the general solution is observed to have the form of simple harmonic motion, as indicated in Eq. (3.160), so that the natural frequency is

$$\omega = \alpha \sqrt{\frac{GJ}{\rho I_p}} \tag{3.171}$$

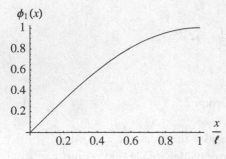

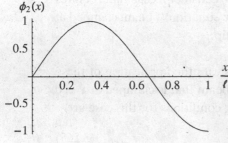

Figure 3.18. First three mode shapes for clamped-free beam vibrating in torsion

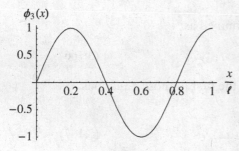

Because α can have only specific values, the frequencies also take on specific numerical values given by

$$\omega_i = \alpha_i \sqrt{\frac{GJ}{\rho I_p}} = \frac{(2i-1)\pi}{2\ell}\sqrt{\frac{GJ}{\rho I_p}} \tag{3.172}$$

These are the natural frequencies of the beam. Associated with each frequency is a "mode shape" as determined from the x-dependent portion of the general solution. The mode shapes (or eigenfunctions) can be written as

$$\phi_i(x) = \sin(\alpha_i x) = \sin\left[\frac{(2i-1)\pi x}{2\ell}\right] \tag{3.173}$$

or any constant times $\phi_i(x)$. The first three of these mode shapes are plotted in Fig. 3.18. The zero derivative at the free end is indicative of the vanishing twisting moment at the free end.

Example Solution for Free-Free Beam. A second example, which exhibits both elastic motion as described previously and motion as a rigid body, is the case of a

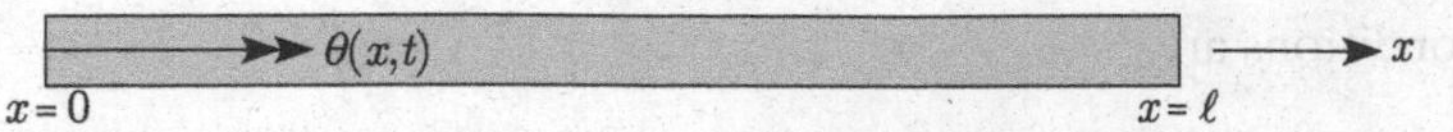

Figure 3.19. Schematic of free-free beam undergoing torsion

beam that is free at both ends, as shown in Fig. 3.19. The boundary conditions are

$$X'(0) = X'(\ell) = 0 \tag{3.174}$$

From the general solution for $X(x)$ in Eqs. (3.138), we find that for $\alpha \neq 0$

$$X'(x) = A\alpha \cos(\alpha x) - B\alpha \sin(\alpha x) \tag{3.175}$$

Thus, the condition at $x = 0$ requires that

$$A\alpha = 0 \tag{3.176}$$

For $A = 0$, the condition at $x = \ell$ requires that

$$\sin(\alpha \ell) = 0 \tag{3.177}$$

because a null solution ($\theta \equiv 0$) is obtained if $B = 0$. This characteristic equation is satisfied by

$$\alpha_i \ell = i\pi \quad (i = 1, 2, \ldots) \tag{3.178}$$

and the corresponding natural frequencies become

$$\omega_i = \frac{i\pi}{\ell} \sqrt{\frac{GJ}{\rho I_p}} \tag{3.179}$$

The associated mode shapes are determined from the corresponding $X(x)$ as

$$\phi_i(x) = \cos(\alpha_i x) = \cos\left(\frac{i\pi x}{\ell}\right) \tag{3.180}$$

These frequencies and mode shapes describe the normal mode of vibration for the elastic degrees of freedom of the free-free beam in torsion.

Now, if in the previous analysis the separation constant, α, is taken as zero, then the governing ordinary differential equations are changed to

$$\frac{X''}{X} = \frac{\rho I_p}{GJ} \frac{\ddot{Y}}{Y} = 0 \tag{3.181}$$

or

$$X''(x) = 0 \text{ and } \ddot{Y}(t) = 0 \tag{3.182}$$

The general solutions to these equations can be written as

$$X(x) = ax + b$$
$$Y(t) = ct + d \tag{3.183}$$

The arbitrary constants, a and b, in the spatially dependent portion of the solution again can be determined from the boundary conditions. For the present case of the free-free beam, the conditions are

$$X'(0) = 0 \text{ requires } a = 0$$

$$X'(\ell) = 0 \text{ requires } a = 0$$

(3.184)

Because both conditions are satisfied without imposing any restrictions on the constant b, this constant can be anything, which implies that the torsional deflection can be nontrivial for $\alpha = 0$. From $X(x)$ with $a = 0$, it is apparent that the corresponding value of θ is independent of the coordinate x. This means that this motion for $\alpha = 0$ is a "rigid-body" rotation of the beam.

The time-dependent solution for this motion, $Y(t)$, also is different from that obtained for the elastic motion. Primarily, the motion is not oscillatory; thus, the rigid-body natural frequency is zero. The arbitrary constants, c and d, can be obtained from the initial values of the rigid-body orientation and angular velocity. To summarize the complete solution for the free-free beam in torsion, a set of generalized coordinates can be defined by

$$\theta(x, t) = \sum_{i=0}^{\infty} \phi_i(x)\xi_i(t)$$

(3.185)

where

$$\phi_0 = 1$$

$$\phi_i = \cos\left(\frac{i\pi x}{\ell}\right) \quad (i = 1, 2, \ldots)$$

(3.186)

The first three elastic mode shapes are plotted in Fig. 3.20. The zero derivative at both ends is indicative of the vanishing twisting moment there. The natural frequencies associated with these mode shapes are

$$\omega_0 = 0$$

$$\omega_i = \frac{i\pi}{\ell}\sqrt{\frac{GJ}{\rho I_p}} \quad (i = 1, 2, \ldots)$$

(3.187)

Note that the rigid-body generalized coordinate, $\xi_0(t)$, represents the radian measure of the rigid-body rotation of the beam about the x axis.

A quick way to verify the existence of a rigid-body mode is to substitute $\omega = 0$ and $X = $ a constant into the differential equation, and boundary conditions for X. A rigid-body mode exists if and only if all are satisfied. Caution: Do *not* try to argue that there is a rigid-body mode because $\alpha = 0$ satisfies the characteristic equation, Eq. (3.177). To obtain that equation, we presupposed that $\alpha \neq 0$!

Example Solution for Clamped-Spring-Restrained Beam. A final example for beam torsion is given by the system shown in Fig. 3.21. The beam is clamped at the root ($x = 0$) end, and the other end is restrained with a rotational spring having spring constant $k = \zeta\overline{GJ}/\ell$, where ζ is a dimensionless parameter. The boundary

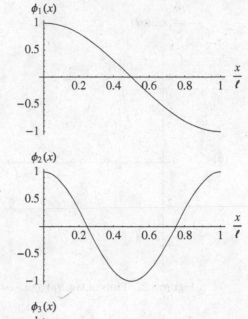

Figure 3.20. First three elastic mode shapes for free-free beam vibrating in torsion

conditions on X are thus

$$X(0) = 0$$

$$\overline{GJ}\,X'(\ell) = -kX(\ell) = -\frac{\overline{GJ}}{\ell}\zeta\,X(\ell) \rightarrow \ell X'(\ell) + \zeta X(\ell) = 0 \tag{3.188}$$

When these boundary conditions are substituted into the general solution found in Eqs. (3.138), we see that the first condition requires that $B = 0$; the second condition, along with the requirement for a nontrivial solution, leads to

$$\zeta \tan(\alpha\ell) + \alpha\ell = 0 \tag{3.189}$$

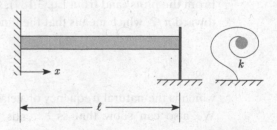

Figure 3.21. Schematic of torsion problem with spring

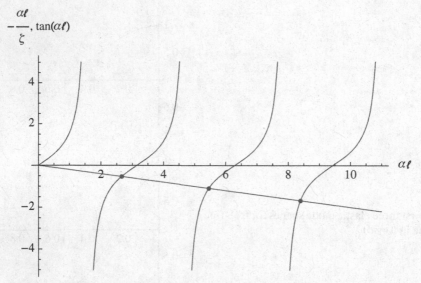

Figure 3.22. Plots of $\tan(\alpha\ell)$ and $-\alpha\ell/\zeta$ versus $\alpha\ell$ for $\zeta = 5$

This transcendental equation has a denumerably infinite set of roots that cannot be found in closed form. However, as many of these roots as desired can be found using numerical procedures found in commercially available software packages such as Mathematica,™ Maple,™ and MATLAB.™

To facilitate this sort of root-finding in general, we may need to specify initial guesses for the values of $\alpha\ell$. These can be found using graphical means by plotting, for example, $\tan(\alpha\ell)$ and $-\alpha\ell/\zeta$ versus $\alpha\ell$ for a specified value of $\zeta = 5$, as shown in Fig. 3.22. The points where these curves intersect (indicated by dots in the figure) are the solutions, the locations of which are seen to be approximately at $\alpha\ell = 2.6$, 5.4, and 8.4. These values, when used as initial guesses in a root-finding application, provide quick convergence to $\alpha_1\ell = 2.65366$, $\alpha_2\ell = 5.45435$, and $\alpha_3\ell = 8.39135$. As an alternative approach for this particular example, we may solve Eq. 3.189 for ζ and plot it versus $\alpha\ell$ to find the roots without iteration.

Thus, the roots of Eq. (3.189) are functions of ζ, and the first four such roots are plotted versus ζ in Fig. 3.23. Denoting these roots by α_i, with $i = 1, 2, \ldots$, we obtain the corresponding natural frequencies

$$\omega_i = \alpha_i \sqrt{\frac{GJ}{\rho I_p}} \qquad (i = 1, 2, \ldots) \tag{3.190}$$

From the plots (and from Eq. 3.189), we note that as ζ tends toward zero, $\alpha_1\ell$ tends toward $\pi/2$, which means that the fundamental natural frequency is

$$\omega_1 = \frac{\pi}{2\ell} \sqrt{\frac{GJ}{\rho I_p}} \qquad (\zeta \to 0) \tag{3.191}$$

which is the natural frequency of a clamped-free beam in torsion (as shown herein). We also can show that as ζ tends to infinity, $\alpha_1\ell$ tends toward π so that the

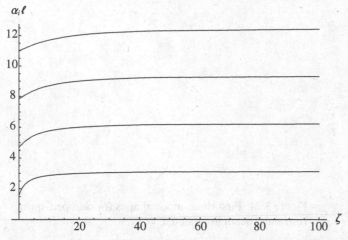

Figure 3.23. Plot of the lowest values of α_i versus ζ for a clamped-spring-restrained beam in torsion

fundamental natural frequency is

$$\omega_1 = \frac{\pi}{\ell} \sqrt{\frac{GJ}{\rho I_p}} \qquad (\zeta \to \infty) \tag{3.192}$$

which is the natural frequency of a clamped-clamped beam in torsion. Recalling the similarity of the governing equations and boundary conditions, the determination of the natural frequencies of a clamped-clamped beam in torsion follows directly from the previous solution for natural frequencies of a string fixed at both ends.

To obtain the corresponding mode shapes, we take the solutions for α_i and substitute back into X, recalling that we can arbitrarily set $A = 1$ and that $B = 0$. The resulting mode shape is

$$\phi_i = \sin(\alpha_i x) \qquad (i = 1, 2, \ldots) \tag{3.193}$$

The first three modes for $\zeta = 1$ are shown in Fig. 3.24. As expected, neither the twist angle nor its derivative are equal to zero at the tip. Close examination of Fig. 3.24 illustrates that for higher and higher frequencies, the spring-restrained ends behave more and more like free ends.

3.2.4 Calculation of Forced Response

The formulation of initial-value problems for beams in torsion is almost identical to that for strings, presented in Section 3.1.7. We first should determine the virtual work done by the applied loads, such as a distributed twisting moment per unit length discussed in Section 2.3.1. From this, we may find the generalized forces associated with torsion. Once the generalized forces are known, we may solve the generalized equations of motion, which are of the form in Eq. (3.90). The resulting initial-value problem then can be solved by invoking orthogonality to obtain values

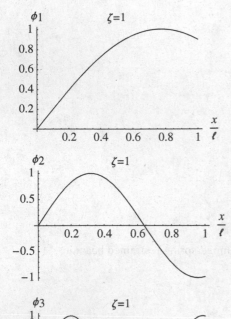

Figure 3.24. First three mode shapes for clamped-spring-restrained beam in torsion, $\zeta = 1$

of the arbitrary constants in the general and particular solutions, as illustrated in the examples in Section 3.1.7.

3.3 Uniform Beam Bending Dynamics

The free vibration of a beam in bending is often referred to as "transverse vibration." This type of motion differs from the transverse dynamics of strings and the torsional dynamics of beams in that the governing equations of motion are of a different mathematical form. Although these equations are different, their solutions are obtained in a similar manner and exhibit similar physical characteristics. Again, we start with the properties varying with x and specialize when we must. Observe that whereas most aerospace structures experience combined or simultaneous bending and torsional dynamic behavior, we have here chosen certain configuration variables to uncouple these types of motion.

3.3.1 Equation of Motion

From Section 2.3.2, Eq. (2.53) is repeated here for convenience

$$\frac{\partial^2}{\partial x^2}\left(\overline{EI}\frac{\partial^2 v}{\partial x^2}\right) + m\frac{\partial^2 v}{\partial t^2} = f(x,t) \tag{3.194}$$

In the following sections, we treat the special case of free vibration for which $f(x,t) = 0$.

3.3.2 General Solutions

A solution to the equation of motion for the transverse vibrations of beams can be obtained by a separation of the independent variables. This separation is denoted as

$$v(x,t) = X(x)Y(t) \tag{3.195}$$

which, when substituted into the equation of motion, yields

$$\frac{(\overline{EI}X'')''}{mX} = -\frac{\ddot{Y}}{Y} \tag{3.196}$$

Because the dependencies on x and t were separated across the equality, each side must equal a constant—say, ω^2. The resulting ordinary differential equations then become

$$(\overline{EI}X'')'' - m\omega^2 X = 0$$
$$\ddot{Y} + \omega^2 Y = 0 \tag{3.197}$$

For simplicity, we specialize the equations for the case of spanwise uniformity of all properties so that the first of Eqs. (3.197) simplifies to

$$X'''' = \alpha^4 X \tag{3.198}$$

where

$$\alpha^4 = \frac{m\omega^2}{\overline{EI}} \tag{3.199}$$

is a constant.

For $\alpha \neq 0$, the general solution to the second of Eqs. (3.197) can be written as in the cases for the string and beam torsion; namely

$$Y(t) = A\sin(\omega t) + B\cos(\omega t) \tag{3.200}$$

For $\alpha \neq 0$, the general solution to the spatially dependent equation can be obtained by presuming a solution of the form

$$X(x) = \exp(\lambda x) \tag{3.201}$$

Substitution of this assumed form into the fourth-order differential equation for $X(x)$ yields

$$\lambda^4 - \alpha^4 = 0 \tag{3.202}$$

which can be factored to

$$(\lambda - i\alpha)(\lambda + i\alpha)(\lambda - \alpha)(\lambda + \alpha) = 0 \tag{3.203}$$

indicating a general solution of the form

$$X(x) = C_1 \exp(i\alpha x) + C_2 \exp(-i\alpha x) + C_3 \exp(\alpha x) + C_4 \exp(-\alpha x) \tag{3.204}$$

Rewriting the exponential functions as trigonometric and hyperbolic sine and cosine functions yields an alternative form of the general solution as

$$X(x) = D_1 \sin(\alpha x) + D_2 \cos(\alpha x) + D_3 \sinh(\alpha x) + D_4 \cosh(\alpha x) \qquad (3.205)$$

Eventual determination of the constants D_i ($i = 1, 2, 3$, and 4) and α requires specification of appropriate boundary conditions. To facilitate this procedure, this last solution form can be rearranged to provide, in some cases, a slight advantage in the algebra, so that

$$\begin{aligned} X(x) = {} & E_1[\sin(\alpha x) + \sinh(\alpha x)] + E_2[\sin(\alpha x) - \sinh(\alpha x)] \\ & + E_3[\cos(\alpha x) + \cosh(\alpha x)] + E_4[\cos(\alpha x) - \cosh(\alpha x)] \end{aligned} \qquad (3.206)$$

To complete the solution, the constants A and B can be determined from the initial deflection and rate of deflection of the beam. The remaining four constants, C_i, D_i, or E_i ($i = 1, 2, 3$, and 4), can be evaluated from the boundary conditions, which must be imposed at each end of the beam. As was true for torsion, the important special case of $\alpha = 0$ is connected with rigid-body modes for beam bending and is addressed in more detail in Section 3.3.4.

3.3.3 Boundary Conditions

For the beam-bending problem, it is necessary to impose two boundary conditions at each end of the beam. Mathematically, boundary conditions may affect v and its partial derivatives, such as $\partial v / \partial x$, $\partial^2 v / \partial x^2$, $\partial^3 v / \partial x^3$, $\partial^2 v / \partial t^2$, and $\partial^3 v / \partial x \partial t^2$. In the context of the separation of variables, these conditions lead to corresponding constraints on some or all of the following at the ends: X, X', X'', and X'''. The resulting four boundary conditions on X and its derivatives are necessary and sufficient for determination of the four constants C_i, D_i, or E_i ($i = 1, 2, 3$, and 4) to within a multiplicative constant.

As with torsion, the nature of the boundary conditions at an end stems from how that end is restrained. When an end cross section is unrestrained, the tractions on it are identically zero. Again, the most stringent condition is a perfect clamp, which for bending allows neither translation nor rotation of an end cross section. Like the clamped-end condition in torsion, the clamped end in bending is a common idealization, although nearly impossible to achieve in practice.

For the bending problem, a wide variety of cases that only partially restrain an end cross section is possible. The cases typically involve elastic and/or inertial reactions. A boundary condition that is idealized in terms of both translational and rotational springs may be used to more realistically account for support flexibility. Appropriate values for both translational and rotational flexibility of the support can be estimated from static tests. Finally, we can use rigid bodies and springs in combination to model attached hardware such as fuel tanks, engines, armaments, and laboratory fixtures.

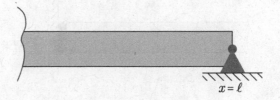

Figure 3.25. Schematic of pinned-end condition

$x = \ell$

In this section, we consider the four "primitive" boundary conditions; four derived boundary conditions involving individual elastic and inertial restraints; and two examples of derived boundary conditions that involve combinations of v and its partial derivatives that can be imposed at the ends of the beam for determination of the four arbitrary constants of the general solution for X.

A boundary condition can be written as a linear relationship involving one or more of the following: the beam deflection, its first three spatial partial derivatives, and its first two temporal partial derivatives. Although it is not a mathematical requirement, the particular combination of conditions to be specified at a beam end should represent a physically realizable constraint. The various spatial partial derivatives of the beam deflection can be associated with particular beam states at any arbitrary point along the beam. There are four such states of practical interest:

1. Deflection $= v(x, t) = X(x)Y(t)$
2. Slope $= \beta(x, t) = \frac{\partial v}{\partial x}(x, t) = X'(x)Y(t)$
3. Bending Moment $= M(x, t) = \overline{EI}(x)\frac{\partial^2 v}{\partial x^2}(x, t) = \overline{EI}(x)X''(x)Y(t)$
4. Shear $= V(x, t) = -\frac{\partial}{\partial x}\left[\overline{EI}(x)\frac{\partial^2 v(x,t)}{\partial x^2}\right] = -[\overline{EI}(x)X''(x)]'Y(t)$

When relating these beam states, the positive convention for deflection and slope is the same at both ends of the beam. In contrast, the sign conventions on shear and bending moment differ at opposite beam ends, as illustrated by the free-body differential beam element used to obtain the equation of motion (see Fig. 2.6).

The most common conditions that can occur at the beam ends involve vanishing pairs of individual states. Typical of such conditions are the following classical configurations (specialized for spanwise uniformity):

- Clamped or built-in end, which implies zero deflection and slope, is illustrated in Fig. 3.10 and has $v(\ell, t) = \frac{\partial v}{\partial x}(\ell, t) = 0$ so that $X(\ell) = X'(\ell) = 0$.
- Free end, which corresponds to zero bending moment and shear, is illustrated in Fig. 3.11 and has $M(\ell, t) = V(\ell, t) = 0$ so that $X''(\ell) = X'''(\ell) = 0$.
- Simply supported, hinged, or pinned end, which indicates zero deflection and bending moment, is denoted by the triangular symbol in Fig. 3.25 and has $v(\ell, t) = M(\ell, t) = 0$ so that $X(\ell) = X''(\ell) = 0$.
- Sliding end, which corresponds to zero shear and slope, is illustrated in Fig. 3.26 and has $\frac{\partial v}{\partial x}(\ell, t) = V(\ell, t) = 0$ so that $X'(\ell) = X'''(\ell) = 0$.

All of these conditions can occur in the same form at $x = 0$.

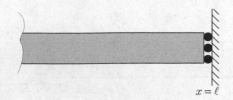

Figure 3.26. Schematic of sliding-end condition

$x = \ell$

In addition to these zero-state conditions, we derive boundary conditions corresponding to linear-constraint reactions associated with elastic and inertial elements. The simplest of these are of the following four basic types:

- translational elastic constraint
- rotational elastic constraint
- translational inertia constraint
- rotational inertia constraint

Two additional examples are presented that are more involved because they entail combinations of these four types.

Translational Elastic Constraint. Consider a beam undergoing bending with a translational spring with elastic constant k attached to the $x = 0$ end of a beam, as shown in Fig. 3.27. Assuming that this end of the beam is deflected by the amount $v(0, t)$, then the spring tries to pull the end of the beam back to its original position by exerting a downward force at the end, the magnitude of which is equal to $kv(0, t)$. Because the transverse-shear force at the left end (on the negative x face) is positive down, the boundary condition becomes

$$V(0, t) = kv(0, t) \tag{3.207}$$

Using the definition of the shear force, we obtain

$$-\frac{\partial}{\partial x}\left(\overline{EI}\frac{\partial^2 v}{\partial x^2}\right)(0, t) = kv(0, t) \tag{3.208}$$

To be useful for separation of variables, we must make the substitution $v(x, t) = X(x)Y(t)$, yielding

$$[\overline{EI}(0)X(0)'']' = -kX(0) \tag{3.209}$$

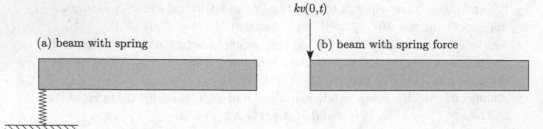

(a) beam with spring

$kv(0,t)$

(b) beam with spring force

Figure 3.27. Example beam undergoing bending with a spring at the $x = 0$ end

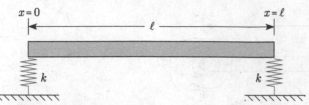

Figure 3.28. Schematic of beam with translational spring at both ends

which further simplifies for a spanwise uniform beam to

$$\overline{EI}X'''(0) = -kX(0) \tag{3.210}$$

If the spring were at $x = \ell$ instead, the direction of the spring force at $x = \ell$ would be the same (i.e., downward), but the shear force is positive upward because this is the positive x face. Thus

$$V(\ell, t) = -kv(\ell, t) \tag{3.211}$$

and

$$[\overline{EI}(\ell)X(\ell)'']' = kX(\ell) \tag{3.212}$$

which further simplifies for a spanwise uniform beam to

$$\overline{EI}X'''(\ell) = kX(\ell) \tag{3.213}$$

These conditions must be augmented by one additional condition at each end because two are required. For example, consider a beam with translational springs at both ends, as shown in Fig. 3.28. At each end, the other conditions for this case would be that the bending moment is equal to zero.

Rotational Elastic Constraint. Consider now a beam with a rotational spring at the right end, as depicted in Fig. 3.29. For a rotation of the end cross-sectional plane of $\partial v/\partial x$ at $x = \ell$, which is positive in the counterclockwise direction, the spring exerts a moment in the opposite direction (i.e., clockwise). Because the bending moment at the right end of the beam is positive in the counterclockwise direction, the boundary condition then becomes

$$M(\ell, t) = -k\frac{\partial v}{\partial x}(\ell, t) \tag{3.214}$$

(a) beam with spring

(b) beam with spring moment

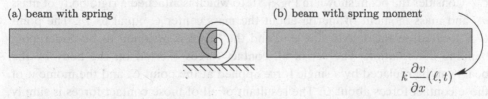

$$k\frac{\partial v}{\partial x}(\ell, t)$$

Figure 3.29. Example of beam undergoing bending with a rotational spring at right end

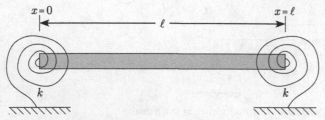

Figure 3.30. Schematic of beam with rotational springs at both ends

Using the definition of bending moment, we find

$$\overline{EI}\frac{\partial^2 v}{\partial x^2}(\ell, t) = -k\frac{\partial v}{\partial x}(\ell, t) \tag{3.215}$$

The boundary condition on X then becomes

$$\overline{EI}X''(\ell) = -kX'(\ell) \tag{3.216}$$

As with the shear force, the sign convention on bending moment differs at the opposite end. Here, at the left end, the spring still exerts a clockwise moment; however, the bending moment is also positive in the clockwise direction. Thus, we may write

$$M(0, t) = k\frac{\partial v}{\partial x}(0, t)$$

$$\overline{EI}\frac{\partial^2 v}{\partial x^2}(0, t) = k\frac{\partial v}{\partial x}(0, t) \tag{3.217}$$

for the condition on $v(x, t)$ and its partial derivatives and

$$\overline{EI}X''(0) = kX'(0) \tag{3.218}$$

for that on $X(x)$ and its derivatives. As before, one more condition is required at each end. Consider, for example, a beam with rotational springs at both ends, shown in Fig. 3.30. Here, it is necessary to set the shear forces at both ends equal to zero.

Translational and Rotational Inertia Constraints. The translational inertia constraint stems from the inertial reaction force associated with the translational motion of either a rigid body or a particle attached to an end of a beam. Similarly, the rotational inertia constraint results from the inertial reaction moment associated with rotational motion of a rigid body attached to an end of a beam.

Consider the beam shown in Fig. 3.31, to which is attached a rigid body of mass m_c and mass moment of inertia about the mass center C equal to I_C. The point C is located on the x axis at $x = 0$, and the beam is assumed to be undergoing bending deformation. The set of all contact forces exerted on the body by the beam can be replaced by a single force applied at the point C, and the moment of those contact forces about C. The resultant of all of those contact forces is simply the shear force $V(0, t)$; their moment about C is the bending moment $M(0, t)$.

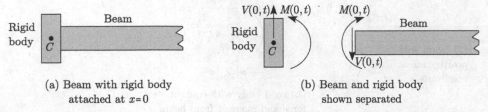

(a) Beam with rigid body
attached at $x=0$

(b) Beam and rigid body
shown separated

Figure 3.31. Schematic of rigid body (a) attached to end of a beam, and (b) detached showing interactions

Therefore, Euler's first and second laws take on the form

$$V(0,t) = m_c \frac{\partial^2 v}{\partial t^2}(0,t)$$

$$M(0,t) = I_C \frac{\partial^3 v}{\partial x \partial t^2}(0,t)$$

(3.219)

which in terms of v and its partial derivatives become, respectively,

$$-\frac{\partial}{\partial x}\left[\overline{EI}(0)\frac{\partial^2 v}{\partial x^2}(0,t)\right] = m_c \frac{\partial^2 v}{\partial t^2}(0,t)$$

$$\overline{EI}(0)\frac{\partial^2 v}{\partial x^2}(0,t) = I_C \frac{\partial^3 v}{\partial x \partial t^2}(0,t)$$

(3.220)

To determine the boundary conditions on X, we first substitute $v(x,t) = X(x)Y(t)$ as before, yielding

$$-[\overline{EI}(0)X''(0)]'Y(t) = m_c X(0)\ddot{Y}(t)$$

$$\overline{EI}(0)X''(0)Y(t) = I_C X'(0)\ddot{Y}(t)$$

(3.221)

Recalling from the second of Eqs. (3.197) that $\ddot{Y} + \omega^2 Y = 0$, these relationships simplify to

$$-[\overline{EI}(0)X''(0)]' = -m_c \omega^2 X(0)$$

$$\overline{EI}(0)X''(0) = -I_C \omega^2 X'(0)$$

(3.222)

which, for a spanwise uniform beam, may be simplified to

$$mX'''(0) = m_c \alpha^4 X(0)$$

$$mX''(0) = -I_C \alpha^4 X'(0)$$

(3.223)

Eqs. (3.223) apply when a rigid body is attached to a free end. For a particle, we may simply set $I_C = 0$. Finally, at the opposite end of the beam, we need only change the signs of the stress resultants, so that

$$-V(\ell,t) = m_c \frac{\partial^2 v}{\partial t^2}(\ell,t)$$

$$-M(\ell,t) = I_C \frac{\partial^3 v}{\partial x \partial t^2}(\ell,t)$$

(3.224)

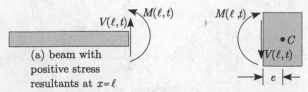

(a) beam with positive stress resultants at $x = \ell$

(b) rigid body with contact force and moment from beam

Figure 3.32. Example with rigid body attached to the right end of beam undergoing bending

from which we may express boundary conditions on X for a spanwise uniform beam, given by

$$mX'''(\ell) = -m_c \alpha^4 X(\ell)$$
$$mX''(\ell) = I_C \alpha^4 X'(\ell) \tag{3.225}$$

It is appropriate to note that the use of the second of Eqs. (3.197) allows us to express Eqs. (3.224) as

$$V(\ell, t) = m_c \omega^2 v(\ell, t)$$
$$M(\ell, t) = I_C \omega^2 \frac{\partial v}{\partial x}(\ell, t) \tag{3.226}$$

subject to the restriction that these conditions hold true only for free vibration.

Other Boundary Configurations. We now turn our attention to two more examples, which are only slightly more involved. First, we consider a beam with an attached rigid body of mass m_c and moment inertia about C given by I_C. The body has a mass center that is offset from the point of attachment (at $x = \ell$) by a distance e, as shown in Fig. 3.32. (This is unlike the previous case in which the body mass center C is located at $x = \ell$ and thus has $e = 0$; see Fig. 3.31.) The body mass center C is assumed to be on the x axis so that transverse vibrations do not excite torsional vibrations and vice versa. The sum of all forces acting on the body is

$$\left(\sum \mathbf{F} \right)_y \equiv -V(\ell, t) \tag{3.227}$$

Euler's first law says that this should be equated to the mass times the acceleration of C. The acceleration of C in the y direction can be written as

$$a_{C_y} = \frac{\partial^2 v}{\partial t^2}(\ell, t) + e \frac{\partial^3 v}{\partial t^2 \partial x}(\ell, t) \tag{3.228}$$

where the body's angular acceleration about the z axis (normal to the plane of the paper) is

$$\alpha_z = \frac{\partial^3 v}{\partial t^2 \partial x}(\ell, t) \tag{3.229}$$

Thus, Euler's first law for the body is

$$-V(\ell, t) = m_c \left[\frac{\partial^2 v}{\partial t^2}(\ell, t) + e \frac{\partial^3 v}{\partial t^2 \partial x}(\ell, t) \right] \tag{3.230}$$

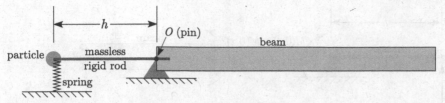

Figure 3.33. Example with mechanism attached to the left end of beam undergoing bending

The sum of the moments about C in the z direction

$$\left(\sum \mathbf{M}_C\right)_z \equiv -M(\ell, t) + eV(\ell, t) \tag{3.231}$$

Euler's second law says that this should be equated to the moment of inertia about C times the angular acceleration about the z axis, so that

$$-M(\ell, t) + eV(\ell, t) = I_C \frac{\partial^3 v}{\partial t^2 \partial x}(\ell, t) \tag{3.232}$$

Eqs. (3.230) and (3.232) can be combined to solve for $V(\ell, t)$ and $M(\ell, t)$, yielding

$$M(\ell, t) = -(I_C + m_c e^2)\frac{\partial^3 v}{\partial t^2 \partial x}(\ell, t) - m_c e \frac{\partial^2 v}{\partial t^2}(\ell, t)$$

$$V(\ell, t) = -m_c \left[\frac{\partial^2 v}{\partial t^2}(\ell, t) + e \frac{\partial^3 v}{\partial t^2 \partial x}(\ell, t)\right] \tag{3.233}$$

where the beam reactions are

$$M(\ell, t) \equiv \overline{EI}(\ell)\frac{\partial^2 v}{\partial x^2}(\ell, t)$$

$$V(\ell, t) \equiv -\frac{\partial}{\partial x}\left[\overline{EI}(\ell)\frac{\partial^2 v}{\partial x^2}(\ell, t)\right] \tag{3.234}$$

The last example involves a mechanism attached to the left end of a beam with a pinned connection to ground, as shown in Fig. 3.33. The massless rigid rod is of length h and the particle has mass m_c; this combination should be considered a rigid body with mass m_c and moment of inertia about the pivot $m_c h^2$. The massless rod is embedded in the left end of the beam and rotates with it. A positive rotation of the $x = 0$ cross-sectional plane about the normal to the page (i.e., the z axis) is counterclockwise and has the value

$$\beta(0, t) = \frac{\partial v}{\partial x}(0, t) \tag{3.235}$$

This rotation results in the downward motion of the particle by the distance $h\beta(0, t)$ and leads to the upward force exerted by the spring, $kh\beta(0, t)$. Thus, this body has a free-body diagram, as shown in Fig. 3.34. A rotation of the rigid body is positive in the counterclockwise direction. Denoting the pivot as O, we find that the sum of moments on the mechanism is

$$\left(\sum \mathbf{M}_O\right)_z \equiv M(0, t) - kh^2 \frac{\partial v}{\partial x}(0, t) \tag{3.236}$$

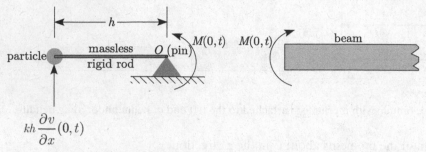

$$kh\frac{\partial v}{\partial x}(0,t)$$

Figure 3.34. Free-body diagram for example with mechanism attached to the left end of beam undergoing bending

Here, Euler's second law is applied about the pivot to avoid dealing with reaction forces at O. This requires us to equate the sum of the moments about O to the moment of inertia about O times the angular acceleration; viz.

$$M(0,t) - kh^2\frac{\partial v}{\partial x}(0,t) = m_c h^2 \frac{\partial^3 v}{\partial t^2 \partial x}(0,t) \tag{3.237}$$

or

$$\overline{EI}\frac{\partial^2 v}{\partial x^2}(0,t) - kh^2\frac{\partial v}{\partial x}(0,t) = m_c h^2 \frac{\partial^3 v}{\partial t^2 \partial x}(0,t) \tag{3.238}$$

The corresponding boundary condition on X at $x = 0$ is found to be

$$\overline{EI}X''(0) - kh^2 X'(0) + m_c h^2 \alpha^4 \frac{\overline{EI}}{m} X'(0) = 0 \tag{3.239}$$

As always with bending problems, one other boundary condition applies at $x = 0$ for the configuration shown in Fig. 3.33—namely, $v(0,t) = X(0) = 0$.

3.3.4 Example Solutions for Mode Shapes and Frequencies

In this section, we consider several examples of the calculation of natural frequencies and mode shapes of vibrating beams in bending. One of the simplest cases is the pinned-pinned case, with which we begin. It is one of the few cases for beams in bending for which a numerical solution of the characteristic equation is *not* required. Next, we treat the important clamped-free case, followed by the case of a hinged-free beam with a rotational restraint about the hinge. Finally, we consider the free-free case, illustrating the concept of the rigid-body mode.

Example Solution for Pinned-Pinned Beam. Consider the pinned-pinned beam as shown in Fig. 3.35. The horizontal rollers at the right end are placed there to indicate that the resultant axial force in the beam is zero. Otherwise, the problem becomes highly nonlinear because it then becomes necessary to take the axial force into account, thereby *significantly* complicating the problem! The boundary conditions reduce to conditions on X given by

$$X(0) = X''(0) = X(\ell) = X''(\ell) = 0 \tag{3.240}$$

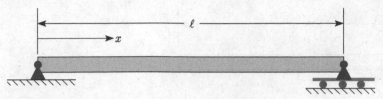

Figure 3.35. Schematic of pinned-pinned beam

Substituting the first two boundary conditions into the general solution as found in Eq. (3.205), we find that

$$D_2 + D_4 = 0$$
$$\alpha^2(-D_2 + D_4) = 0$$

(3.241)

Recall that the constant α cannot be zero. To consider the $\alpha = 0$ case we must take a solution in the form of a cubic polynomial, and the boundary conditions for this case do not yield a nontrivial solution of that form. Therefore, $D_2 = D_4 = 0$, and the solution for X becomes

$$X(x) = D_1 \sin(\alpha x) + D_3 \sinh(\alpha x)$$

(3.242)

Using the last two of the boundary conditions, we obtain a set of homogeneous algebraic equations in D_1 and D_3

$$\begin{bmatrix} \sin(\alpha\ell) & \sinh(\alpha\ell) \\ -\sin(\alpha\ell) & \sinh(\alpha\ell) \end{bmatrix} \begin{Bmatrix} D_1 \\ D_3 \end{Bmatrix} = \begin{Bmatrix} 0 \\ 0 \end{Bmatrix}$$

(3.243)

A nontrivial solution can exist only if the determinant of the coefficients is equal to zero; therefore

$$2 \sin(\alpha\ell) \sinh(\alpha\ell) = 0$$

(3.244)

Because $\alpha \neq 0$, we know that the only way this characteristic equation can be satisfied is for

$$\sin(\alpha\ell) = 0$$

(3.245)

which has a denumerably infinite set of roots given by

$$\alpha_i = \frac{i\pi}{\ell} \qquad (i = 1, 2, \ldots)$$

(3.246)

Although this is the same set of eigenvalues that we found for the string problem, the relationship to the natural frequencies is quite different; viz.

$$\omega_i^2 = \frac{\overline{EI}\alpha_i^4}{m}$$

(3.247)

so that

$$\omega_i = \alpha_i^2 \sqrt{\frac{EI}{m}} = \left(\frac{i\pi}{\ell}\right)^2 \sqrt{\frac{EI}{m}} = (i\pi)^2 \sqrt{\frac{EI}{m\ell^4}}$$

(3.248)

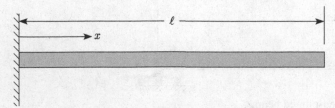

Figure 3.36. Schematic of clamped-free beam

As observed in the cases of the string and beam torsion, there is associated with the ith natural frequency a unique deformation shape called the mode shape (or eigenfunction). Each mode shape can be obtained from the spatially dependent portion of the solution by evaluating the function, $X_i(x)$, for any known value of α_i. To find X_i, we substitute any value for α_i back into either of the two scalar equations represented by the matrix equation in Eq. (3.243). It is important to recognize that the constants D_1 and D_3 now should be written as D_{1i} and D_{3i}. Using the first of these equations along with the knowledge that $\sinh(\alpha_i\ell) \neq 0$, we find that $D_{3i} = 0$, leaving

$$X_i = D_{1i} \sin\left(\frac{i\pi x}{\ell}\right) \qquad (i = 1, 2, \ldots) \tag{3.249}$$

where D_{1i} can be any nonzero constant. For example, choosing $D_{1i} = 1$, we find the mode shape to be

$$\phi_i = \sin\left(\frac{i\pi x}{\ell}\right) \qquad (i = 1, 2, \ldots) \tag{3.250}$$

which is the same mode shape as obtained previously for the vibrating string.

Example Solution for Clamped-Free Beam. Consider the clamped-free beam as shown in Fig. 3.36, the boundary conditions of which reduce to conditions on X given by

$$X(0) = X'(0) = X''(\ell) = X'''(\ell) = 0 \tag{3.251}$$

As in the previous example, we can show that this problem does not exhibit a nontrivial solution for the case of $\alpha = 0$. Thus, we use the form of the general solution in Eq. (3.206) for which $\alpha \neq 0$. Along with the first two boundary conditions, this yields

$$\begin{aligned} X(0) = 0 &\rightarrow E_3 = 0 \\ X'(0) = 0 &\rightarrow E_1 = 0 \end{aligned} \tag{3.252}$$

The remaining boundary conditions yield two homogeneous algebraic equations that may be reduced to the form

$$\begin{bmatrix} \sinh(\alpha\ell) + \sin(\alpha\ell) & \cosh(\alpha\ell) + \cos(\alpha\ell) \\ \cosh(\alpha\ell) + \cos(\alpha\ell) & \sinh(\alpha\ell) - \sin(\alpha\ell) \end{bmatrix} \begin{Bmatrix} E_2 \\ E_4 \end{Bmatrix} = \begin{Bmatrix} 0 \\ 0 \end{Bmatrix} \tag{3.253}$$

Table 3.1. *Values of $\alpha_i \ell$, $(2i - 1)\pi/2$, and β_i for $i = 1, \ldots, 5$ for the clamped-free beam*

i	$\alpha_i \ell$	$(2i - 1)\pi/2$	β_i
1	1.87510	1.57080	0.734096
2	4.69409	4.71239	1.01847
3	7.85476	7.85398	0.999224
4	10.9955	10.9956	1.00003
5	14.1372	14.1372	0.999999

It can be verified by applying Cramer's method for their solution that a nontrivial solution exists only if the determinant of the coefficients is equal to zero. This is typical of all nontrivial solutions to homogeneous, linear, algebraic equations, and here yields

$$\sinh^2(\alpha\ell) - \sin^2(\alpha\ell) - [\cosh(\alpha\ell) + \cos(\alpha\ell)]^2 = 0 \tag{3.254}$$

or, noting the identities

$$\begin{aligned} \sin^2(\alpha\ell) + \cos^2(\alpha\ell) &= 1 \\ \cosh^2(\alpha\ell) - \sinh^2(\alpha\ell) &= 1 \end{aligned} \tag{3.255}$$

we obtain the characteristic equation as simply

$$\cos(\alpha\ell)\cosh(\alpha\ell) + 1 = 0 \tag{3.256}$$

We cannot extract a closed-form exact solution for this transcendental equation. However, numerical solutions are obtained easily. Most numerical procedures require initial estimates of the solution to converge. Because $\cosh(\alpha\ell)$ becomes large as its argument becomes large, we can argue that at least the largest roots will be close to those of $\cos(\alpha\ell) = 0$, or $\alpha_i\ell = (2i - 1)\pi/2$. Indeed, the use of these values as initial estimates yields a set of numerical values that approach the initial estimates ever more closely as i increases. The values of $\alpha_i\ell$ (i.e., dimensionless quantities) are listed in Table 3.1. To six places, all values of $\alpha_i\ell$ for $i \geq 5$ are equal to $(2i - 1)\pi/2$. The corresponding natural frequencies are given by

$$\omega_i = \alpha_i^2 \sqrt{\frac{EI}{m}} = (\alpha_i\ell)^2 \sqrt{\frac{EI}{m\ell^4}} \tag{3.257}$$

To obtain the mode shapes, we substitute the values in Table 3.1 into either of Eqs. (3.253). The resulting equation for the ith mode has one arbitrary constant remaining (i.e., either E_{2i} or E_{4i} can be kept), which can be set equal to any number desired to conveniently normalize the resulting mode shape ϕ_i. For example, normalizing the solution by $-E_{4i}$, which is equivalent to setting $E_{4i} = -1$, we can show that

$$\phi_i = \cosh(\alpha_i x) - \cos(\alpha_i x) - \beta_i[\sinh(\alpha_i x) - \sin(\alpha_i x)] \tag{3.258}$$

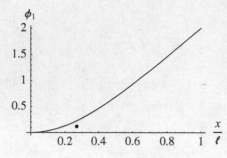

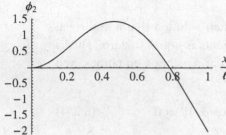

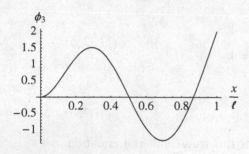

Figure 3.37. First three free-vibration mode shapes of a clamped-free beam in bending

where

$$\beta_i = -\frac{E_{2i}}{E_{4i}} = \frac{\cosh(\alpha_i\ell) + \cos(\alpha_i\ell)}{\sinh(\alpha_i\ell) + \sin(\alpha_i\ell)} \qquad (3.259)$$

The values of β_i also are tabulated in Table 3.1. For this particular normalization

$$\int_0^\ell \phi_i^2\, dx = \ell$$

$$\phi_i(\ell) = 2(-1)^{i+1} \qquad (3.260)$$

the first of which is left to the reader to show (see Prob. 10d). The first three mode shapes are depicted in Fig. 3.37. Note that as with previous results, the higher the mode number, the more nodes (i.e., crossings of the zero-displacement line).

Example Solution for Spring-Restrained, Hinged-Free Beam. This sample problem for which modes of vibration are determined is for a uniform beam that is hinged at the right-hand end and restrained there by a rotational spring with elastic

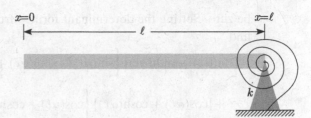

Figure 3.38. Schematic of spring-restrained, hinged-free beam

constant $k = \kappa \overline{EI}/\ell$. The left-hand end is free, as illustrated in Fig. 3.38. The boundary conditions for this case require that

$$X''(0) = 0$$
$$X'''(0) = 0$$
$$X(\ell) = 0 \qquad (3.261)$$
$$\overline{EI}X''(\ell) = -kX'(\ell) \text{ or } \ell X''(\ell) = -\kappa X'(\ell)$$

The spatially dependent portion of the general solution is used in the form of Eq. (3.206). The two conditions of zero bending moment and shear at $x = 0$ require that

$$X''(0) = 0 \rightarrow E_4 = 0$$
$$X'''(0) = 0 \rightarrow E_2 = 0 \qquad (3.262)$$

The third boundary condition, that of zero displacement at $x = \ell$, can now be indicated by

$$X(\ell) = E_1 \left[\sin(\alpha\ell) + \sinh(\alpha\ell) \right] + E_3 \left[\cos(\alpha\ell) + \cosh(\alpha\ell) \right] = 0 \qquad (3.263)$$

The fourth boundary condition, a rotational elastic constraint at $x = \ell$, can be written as

$$\ell^2 X''(\ell) + \kappa \ell X'(\ell) = 0 \qquad (3.264)$$

so that

$$(\alpha\ell)^2 \left\{ E_1 \left[-\sin(\alpha\ell) + \sinh(\alpha\ell) \right] + E_3 \left[-\cos(\alpha\ell) + \cosh(\alpha\ell) \right] \right\}$$
$$+ \kappa\alpha\ell \left\{ E_1 \left[\cos(\alpha\ell) + \cosh(\alpha\ell) \right] + E_3 \left[-\sin(\alpha\ell) + \sinh(\alpha\ell) \right] \right\} = 0 \qquad (3.265)$$

This relationship can be rearranged as

$$E_1 \left\{ \cos(\alpha\ell) + \cosh(\alpha\ell) + \frac{\alpha\ell}{\kappa} \left[-\sin(\alpha\ell) + \sinh(\alpha\ell) \right] \right\}$$
$$+ E_3 \left\{ -\sin(\alpha\ell) + \sinh(\alpha\ell) + \frac{\alpha\ell}{\kappa} \left[-\cos(\alpha\ell) + \cosh(\alpha\ell) \right] \right\} = 0 \qquad (3.266)$$

The simultaneous solution of Eqs. (3.263) and (3.266) for nonzero values of E_1 and E_3 requires that the determinant of the 2×2 array formed by their coefficients must be zero. Setting the determinant formed from Eqs. (3.263) and (3.266) to zero, we find

$$[\sin(\alpha\ell) + \sinh(\alpha\ell)] \left\{ \sin(\alpha\ell) - \sinh(\alpha\ell) + \frac{\alpha\ell}{\kappa} [\cos(\alpha\ell) - \cosh(\alpha\ell)] \right\}$$

$$+ [\cos(\alpha\ell) + \cosh(\alpha\ell)] \left\{ \cos(\alpha\ell) + \cosh(\alpha\ell) + \frac{\alpha\ell}{\kappa} [-\sin(\alpha\ell) + \sinh(\alpha\ell)] \right\} = 0$$

$$(3.267)$$

After executing the indicated multiplications and applying the identities of Eqs. (3.255), the relationship becomes

$$\left(\frac{\alpha\ell}{\kappa} \right) [\sin(\alpha\ell)\cosh(\alpha\ell) - \cos(\alpha\ell)\sinh(\alpha\ell)] = 1 + \cos(\alpha\ell)\cosh(\alpha\ell) \qquad (3.268)$$

This is the characteristic equation. As in the previous example, it is a transcendental equation that cannot be solved analytically. Note that for specified finite and nonzero values of κ, we may calculate numerically a denumerably infinite set of the eigenvalues $\alpha_i\ell$ (for $i = 1, 2, \ldots$) by a suitable iterative procedure. For such an iterative solution, we need initial estimates for the $\alpha\ell$s. Note, however, that this equation is a special case in which we may solve for κ as a function of $\alpha\ell$ without iteration.

In the limit as κ tends to infinity, we find eigenvalues in agreement with the clamped-free case, as expected. In the limit as κ tends to zero, we can show that a rigid-body mode exists. The next example illustrates a procedure by which we may prove the existence of one or more rigid-body modes. It is important to note, however, that it is incorrect to try to infer the existence of a rigid-body mode because $\alpha\ell = 0$ satisfies Eq. (3.268) in the limit as κ tends to zero; our general solution for X is valid only when $\alpha \neq 0$.

For specified values of m, $\overline{EI}$, ℓ, and the stiffness parameter κ, the eigenvalues can be used to determine the natural frequencies as

$$\omega_i = \alpha_i^2 \sqrt{\frac{EI}{m}} = (\alpha_i\ell)^2 \sqrt{\frac{EI}{m\ell^4}} \quad (i = 1, 2, \ldots) \qquad (3.269)$$

and the ith mode shape can be defined as

$$\phi_i(x) = \frac{X_i(x)}{E_{1i}}$$

$$= \sin(\alpha_i x) + \sinh(\alpha_i x) + \beta_i [\cos(\alpha_i x) + \cosh(\alpha_i x)] \qquad (3.270)$$

The modal parameter $\beta_i = E_{3i}/E_{1i}$ can be obtained from the zero-displacement boundary condition at $x = \ell$, Eq. (3.263). When evaluated for the ith mode, β_i becomes

$$\beta_i = \frac{E_{3i}}{E_{1i}} = -\frac{\sin(\alpha_i\ell) + \sinh(\alpha_i\ell)}{\cos(\alpha_i\ell) + \cosh(\alpha_i\ell)} \qquad (3.271)$$

numerical values of which can be found once $\alpha_i\ell$ is known for specific values of κ.

A sample set of numerical results for this example is shown in Figs. 3.39 through 3.41. The first three mode shapes are shown for $\kappa = 1$ in Fig. 3.39. Fig. 3.40 shows the variation of $\alpha_i\ell$ versus κ for $i = 1, 2,$ and 3, illustrating the fact that the frequencies of

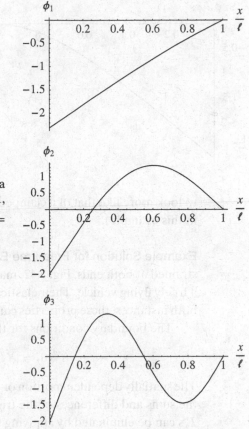

Figure 3.39. Mode shapes for first three modes of a spring-restrained, hinged-free beam in bending; $\kappa = 1$, $\omega_1 = (1.24792)^2 \sqrt{EI/(m\ell^4)}$, $\omega_2 = (4.03114)^2 \sqrt{EI/(m\ell^4)}$, and $\omega_3 = (7.13413)^2 \sqrt{EI/(m\ell^4)}$

the higher modes are much less sensitive to the spring constant than that of the first mode. Indeed, the first mode frequency (proportional to the square of the smallest plotted quantity in Fig. 3.40) tends to zero as κ tends toward zero in the limit. This can be interpreted as the lowest-frequency mode transitioning to a rigid-body mode, which exists only when the spring constant is identically zero. In the limit as κ becomes infinite, in contrast, the eigenvalues tend toward those of the clamped-free beam, as expected. Indeed, as Fig. 3.41 shows, when $\kappa = 50$ the mode shape starts

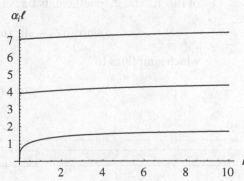

Figure 3.40. Variation of lowest eigenvalues $\alpha_i \ell$ versus dimensionless spring constant κ

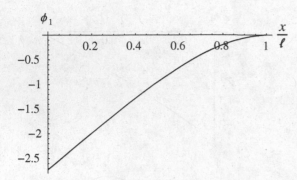

Figure 3.41. Mode shape for fundamental mode of the spring-restrained, hinged-free beam in bending; $\kappa = 50$, $\omega_1 = (1.83929)^2 \sqrt{EI/(m\ell^4)}$

to look more like that of a clamped-free beam (with the fixity being on the right end in this example).

Example Solution for Free-Free Beam. The case of a uniform beam that is unconstrained at both ends, Fig. 3.42, may be considered as a crude first approximation to a freely flying vehicle. Their elastic and rigid dynamic properties are quite similar. In both instances, these properties can be described in terms of a modal representation.

The boundary conditions for this case require that

$$X''(0) = X'''(0) = X''(\ell) = X'''(\ell) = 0 \qquad (3.272)$$

The spatially dependent portion of the general solution to be used here again involves the sums and differences of the trigonometric and hyperbolic functions. Two of the E_is can be eliminated by applying the boundary conditions at $x = 0$ so that

$$X''(0) = 0 \to E_4 = 0$$
$$X'''(0) = 0 \to E_2 = 0 \qquad (3.273)$$

The conditions at $x = \ell$ of zero bending moment and zero shear $X''(\ell) = 0$, and $X'''(\ell) = 0$, respectively, yield the following relationships:

$$E_1 \left[-\sin(\alpha\ell) + \sinh(\alpha\ell) \right] + E_3 \left[-\cos(\alpha\ell) + \cosh(\alpha\ell) \right] = 0$$
$$E_1 \left[-\cos(\alpha\ell) + \cosh(\alpha\ell) \right] + E_3 \left[\sin(\alpha\ell) + \sinh(\alpha\ell) \right] = 0 \qquad (3.274)$$

Here again, the nontrivial solution to these equations requires that the determinant of the E_1 and E_3 coefficients be zero. This relationship becomes

$$\sinh^2(\alpha\ell) - \sin^2(\alpha\ell) - [\cosh(\alpha\ell) - \cos(\alpha\ell)]^2 = 0 \qquad (3.275)$$

which simplifies to

$$\cos(\alpha\ell)\cosh(\alpha\ell) = 1 \qquad (3.276)$$

$x = 0$ $x = \ell$

Figure 3.42. Schematic of free-free beam

Table 3.2. *Values of $\alpha_i \ell$, $(2i + 1)\pi/2$, and β_i for*
$i = 1, \ldots, 5$ for the free-free beam

i	$\alpha_i \ell$	$(2i+1)\pi/2$	β_i
1	4.73004	4.71239	0.982502
2	7.85320	7.85398	1.00078
3	10.9956	10.9956	0.999966
4	14.1372	14.1372	1.00000
5	17.2788	17.2788	1.00000

For large $\alpha\ell$, the roots tend to values that make $\cos(\alpha\ell) = 0$. Unlike the clamped-free case, however, there is no root near $\pi/2$, and the first nonzero root occurs near $3\pi/2$. Indeed, the ith root is near $(2i + 1)\pi/2$. Thus, the roots of this characteristic equation readily can be computed numerically to yield the eigenvalues $\alpha_i \ell$ in Table 3.2. From these numerical values, the natural frequencies can be found as

$$\omega_i = \alpha_i^2 \sqrt{\frac{EI}{m}} \tag{3.277}$$

The mode shape associated with each eigenvalue can be defined as

$$\phi_i(x) = \frac{X_i(x)}{E_{3i}} \tag{3.278}$$

$$= \cos(\alpha_i x) + \cosh(\alpha_i x) - \beta_i \left[\sin(\alpha_i x) + \sinh(\alpha_i x)\right]$$

The numerical value of the modal parameter $\beta_i = -E_{1i}/E_{3i}$, also tabulated in Table 3.2, can be obtained from either of the boundary conditions given in Eqs. (3.274). Using the first of those equations as an example, we obtain

$$\beta_i = -\frac{E_{1i}}{E_{3i}} = \frac{\cosh(\alpha_i\ell) - \cos(\alpha_i\ell)}{\sinh(\alpha_i\ell) - \sin(\alpha_i\ell)} \tag{3.279}$$

It can be shown that the first of Eqs. (3.274) would yield the same result by using the characteristic equation as an identity. The first three of these mode shapes are shown in Fig. 3.43.

In addition to these modal properties that can be used to describe the elastic behavior of the beam, there are also modal properties that describe the rigid behavior of the beam. These modes are associated with zero values of the separation constant α. Recall that a similar result was obtained for torsional deflections of a free-free beam. When α is zero, the governing ordinary differential equations for beam bending, Eqs. (3.197), become

$$X'''' = 0 \quad \ddot{Y} = 0 \tag{3.280}$$

The general solutions to these equations can be written as

$$X = \frac{bx^3}{6} + \frac{cx^2}{2} + dx + e \tag{3.281}$$

$$Y = ft + g$$

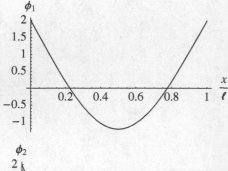

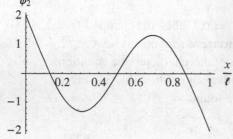

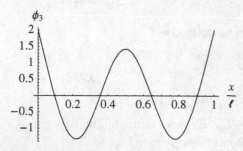

Figure 3.43. First three free-vibration elastic mode shapes of a free-free beam in bending

where the arbitrary constants, b through e, in the spatially dependent portion of the solution can be established from the boundary conditions. These conditions of zero bending moment and shear at the ends of the beam yield the following:

$$X''(0) = 0 \rightarrow c = 0$$

$$X'''(0) = 0 \rightarrow b = 0$$

$$X''(\ell) = 0 \rightarrow b\ell + c = 0 \qquad (3.282)$$

$$X'''(\ell) = 0 \rightarrow b = 0$$

It is apparent that all four boundary conditions can be satisfied with $b = c = 0$. Because no restrictions are placed on the constants d and e, they can be arbitrary. Thus, a general description of the solution in this case is

$$X = dx + e \qquad (3.283)$$

An important characteristic of this solution is that no relationship has been established between d and e. Therefore, they can be presumed to represent two independent motions of the beam. As written previously, e represents a rigid vertical translation of the beam because it is independent of x. The dx term, being linear in x, represents a rigid rotation of the beam about the left end. It can be shown that when the rotational motion is taken to be about the mass centroid, it and the translation are orthogonal with respect to one another and with respect to the elastic modes. This suggests that the modal representation for these rigid-body degrees of freedom can be described by

$$v_{\text{rigid}} = \sum_{i=-1}^{0} \phi_i(x)\xi_i(t) \tag{3.284}$$

where

$$\phi_{-1} = 1 \quad \text{and} \quad \xi_{-1}(t) = \text{translation}$$
$$\phi_0 = x - \frac{\ell}{2} \quad \text{and} \quad \xi_0(t) = \text{rotation angle} \tag{3.285}$$

The time-dependent portion of the solution for these rigid-body motions is seen to be aperiodic. This means that natural frequencies for both rigid-body modes are zero. The two arbitrary constants contained in $Y(t)$ can be evaluated from the initial rigid-body displacement and velocity associated with the translation and rotation. Thus, the complete solution for the free-free beam-bending problem can now be written in terms of all of its modes as

$$v = \sum_{i=-1}^{\infty} \phi_i(x)\xi_i(t) \tag{3.286}$$

This example provides a convenient vehicle for further discussion of symmetry. It was already noted in the case of a vibrating string that systems exhibiting geometric symmetry have two distinct types of mode shapes—namely, those that are symmetric about the midpoint and those that are antisymmetric about the midpoint. As can be seen in the results, this is indeed true for the modes of the free-free beam. In particular, the rigid-body translation mode and the first and third elastic modes are clearly symmetric about the midpoint of the beam, whereas the rigid-body rotation mode and the second elastic mode are antisymmetric about the midpoint (see Fig. 3.43).

This observation suggests that the symmetric mode shapes could be obtained by calculating the mode shapes of a beam that is half the length of the original beam and that has the sliding condition at one end and is free at the other. Similarly, the antisymmetric modes could be obtained by calculating the mode shapes of a beam with half the length of the original beam and that has one end pinned and the other free. It also should be evident that a symmetric aircraft with high-aspect-ratio wings, modeled as beams and attached to a rigid-body fuselage, could be represented similarly in terms of the symmetric and antisymmetric modes of the combined body and wing system. That is, we may model the whole system by considering only one wing attached to a rigid body with half the mass and half the rotational inertia with appropriate boundary conditions.

3.3.5 Calculation of Forced Response

The formulation of initial-value problems for beams in bending is almost identical to that for beams in torsion and for strings; see Sections 3.1.7 and 3.2.4, respectively. We should first determine the virtual work done by the applied loads, such as a distributed transverse force per unit length. From this, we may find the generalized forces associated with bending. Once they are known, one may solve the generalized equations of motion, which are in the form of Eq. (3.90). The resulting initial-value problem then can be solved by invoking orthogonality to obtain values of the arbitrary constants in the general and particular solutions, as illustrated in the examples in Section 3.1.7.

3.4 Free Vibration of Beams in Coupled Bending and Torsion

In this section, the analytical treatment of coupled bending–torsion vibration of composite beams is briefly considered. The treatment is restricted to uniform beams and to the presentation of governing equations, sample boundary conditions, and suggestions for solution.

3.4.1 Equations of Motion

First, we specialize Eqs. (2.65) for spanwise uniformity and free vibration, yielding

$$\overline{\rho I_p}\frac{\partial^2 \theta}{\partial t^2} + md\frac{\partial^2 v}{\partial t^2} - \overline{GJ}\frac{\partial^2 \theta}{\partial x^2} + K\frac{\partial^3 v}{\partial x^3} = 0$$

$$md\frac{\partial^2 \theta}{\partial t^2} + m\frac{\partial^2 v}{\partial t^2} + \overline{EI}\frac{\partial^4 v}{\partial x^4} - K\frac{\partial^3 \theta}{\partial x^3} = 0 \tag{3.287}$$

Because these are linear equations with constant coefficients, for free vibration we may assume simple harmonic motion. In the spirit of separation of variables, the solutions for v and θ are written as

$$v(x, t) = \overline{v}(x)\exp(i\omega t)$$

$$\theta(x, t) = \overline{\theta}(x)\exp(i\omega t) \tag{3.288}$$

with the mode shapes being of the form

$$\overline{v} = \hat{v}\exp(\alpha x)$$

$$\overline{\theta} = \hat{\theta}\exp(\alpha x) \tag{3.289}$$

which allows us to write the system of equations in matrix form as

$$\begin{bmatrix} \overline{EI}\alpha^4 - m\omega^2 & -K\alpha^3 - md\omega^2 \\ K\alpha^3 - md\omega^2 & -\overline{GJ}\alpha^2 - \overline{\rho I_p}\omega^2 \end{bmatrix}\begin{Bmatrix} \hat{v} \\ \hat{\theta} \end{Bmatrix} = \begin{Bmatrix} 0 \\ 0 \end{Bmatrix} \tag{3.290}$$

For there to be a nontrivial solution, the determinant of the coefficient matrix must vanish, yielding

$$(\overline{EI}\,\overline{GJ} - K^2)\alpha^6 + \overline{\rho I_p}\,\overline{EI}\omega^2\alpha^4 - m\overline{GJ}\omega^2\alpha^2 - (m\overline{\rho I_p} - m^2 d^2)\omega^4 = 0 \quad (3.291)$$

This cubic equation in α^2 may be solved for arbitrary ω^2. When d and K are nonzero, finding the exact, closed-form solution "by hand" is problematic. However, with the aid of symbolic computational tools such as Mathematica™, we may easily extract the six roots denoted here by α_i for $i = 1, 2, \ldots, 6$, as functions of ω^2. Note that $\alpha_{i+3} = -\alpha_i$ for $i = 1, 2$, and 3.

Therefore, when $K, d \neq 0$, the solution for the mode shape may be written as

$$\begin{aligned}
\bar{v} &= C_1 \exp(\alpha_1 x) + C_2 \exp(\alpha_2 x) + C_3 \exp(\alpha_3 x) \\
&\quad + C_4 \exp(-\alpha_1 x) + C_5 \exp(-\alpha_2 x) + C_6 \exp(-\alpha_3 x) \\
\bar{\theta} &= D_1 \exp(\alpha_1 x) + D_2 \exp(\alpha_2 x) + D_3 \exp(\alpha_3 x) \\
&\quad + D_4 \exp(-\alpha_1 x) + D_5 \exp(-\alpha_2 x) + D_6 \exp(-\alpha_3 x)
\end{aligned} \quad (3.292)$$

where

$$D_i = C_i \left(\frac{K\alpha_i^3 - md\omega^2}{\overline{GJ}\alpha_i^2 + \overline{\rho I_p}\omega^2} \right) \quad i = 1, 2, \ldots, 6 \quad (3.293)$$

Now, with six boundary conditions (i.e., three at each end), we may find six homogeneous algebraic equations for C_i. The condition for a nontrivial solution leads to the characteristic equation for ω^2. There is a denumerably infinite set of roots for ω^2, so that for any value determined for ω^2, we may find any five of the C_i coefficients in terms of the sixth and thus determine the mode shapes. In general, each mode shape involves both v and θ. For small couplings (i.e., such that $K^2 \ll \overline{GJ}\,\overline{EI}$ and $md^2 \ll \overline{\rho I_p}$), one "branch" of these roots is near the uncoupled bending frequencies and the other is near the uncoupled torsional frequencies.

3.4.2 Boundary Conditions

The boundary conditions for coupled bending and torsion range from very simple to somewhat complex, depending on the type of restraint(s) imposed on the ends. For example, for a clamped end, we have $v = \partial v/\partial x = \theta = 0$, the same as for uncoupled bending and torsion. Similarly, a free end has zero bending moment, shear force, and twisting moment, respectively written as $M = V = T = 0$. Note the definitions of M and T in Eqs. (2.58) and that $V = -\partial M/\partial x$. Equations governing other restraints may be determined by appropriate kinematical or physical relationships. For example, a pinned connection may imply specification of an axis about which the moment vector (i.e., combination of bending and twisting moments) vanishes and perpendicular to which components of the rotation vector (i.e., combination of bending and twisting rotations) vanish. Relationships for both elastic and inertial restraints may be developed using Euler's laws, as in the uncoupled cases herein.

The complexity of this class of problem provides excellent motivation for the introduction of approximate methods, which is undertaken in the next section.

3.5 Approximate Solution Techniques

There are several popular methods that make use of a set of modes or other functions to approximate the dynamic behavior of systems. In this section, without going into detail about the theories associated with this subject, we illustrate within the framework already established how we can use a truncated set of modes or another set of functions to obtain an approximate solution. Details of the theories behind modal approximation methods are found in texts that treat structural dynamics at the graduate level. The two main approaches are (1) Galerkin's method, applied to ordinary or partial differential equations; and (2) the Ritz method, applied to Lagrange's equations or the principle of virtual work. These two methods yield identical results in certain situations. Thus, if time is limited, it would be necessary to discuss only one of the two methods to give students an introduction to the method and an appreciation of results that can be obtained this way. The Ritz method is preferred in the present context because of the ease with which it can be presented within the framework of Lagrange's equations. Nevertheless, both of these methods are presented at a level suitable for undergraduate students.

3.5.1 The Ritz Method

Building on the previous treatment, we start with Lagrange's equations, given by

$$\frac{d}{dt}\left(\frac{\partial L}{\partial \dot{\xi}_i}\right) - \frac{\partial L}{\partial \xi_i} = \Xi_i \qquad i = 1, 2, \ldots, n \tag{3.294}$$

where in the Lagrangean, $L = K - P$, the total kinetic energy is K, the total potential energy is P, n is the number of generalized coordinates retained, the generalized coordinates are ξ_i, and Ξ_i is the generalized force. Although it can be helpful, as discussed herein, it is not necessary to make use of potential energy, which can account only for conservative forces. The generalized force, however, can be used to include the effects of *any* loads. So as not to count the same physical effects more than once, the generalized force should include only those forces that are not accounted for in the potential energy. The generalized forces stem from virtual work, which can be written as

$$\overline{\delta W} = \sum_{i=1}^{n} \Xi_i \delta \xi_i \tag{3.295}$$

where $\delta \xi_i$ is an arbitrary increment in the ith generalized coordinate.

Consider a beam in bending as an example. The total kinetic energy must include that of the beam as well as any attached particles or rigid bodies. The contribution of the beam is

$$K_{\text{beam}} = \frac{1}{2} \int_0^\ell m \left(\frac{\partial v}{\partial t} \right)^2 dx \qquad (3.296)$$

where m is the mass per unit length of the beam. The total potential energy $P = U + V$ comprises the internal strain energy of the beam, denoted by U, plus any additional potential energy, V, attributed to gravity, springs attached to the beam, or applied static loads. All other loads, such as aerodynamic loads, damping, and follower forces, must be accounted for in Ξ_i.

The strain energy for a beam in bending is given by

$$U = \frac{1}{2} \int_0^\ell \overline{EI} \left(\frac{\partial^2 v}{\partial x^2} \right)^2 dx \qquad (3.297)$$

The expression for V varies depending on the problem being addressed, as does the virtual work of all forces other than those accounted for in V. The virtual work of an applied distributed force per unit length $f(x, t)$ can be written as

$$\overline{\delta W} = \int_0^\ell f(x, t) \delta v(x, t) dx \qquad (3.298)$$

where δv is an increment of v in which time is held fixed and $f(x, t)$ is positive in the direction of positive v.

To apply the Ritz method, we need to express P, K, and $\overline{\delta W}$ in terms of a series of functions with one or more terms. For a beam in bending, this means that

$$v(x, t) = \sum_{i=1}^n \xi_i(t) \phi_i(x) \qquad (3.299)$$

There are several characteristics that these "basis functions" ϕ_i must possess, as follows:

1. Each function must satisfy at least all boundary conditions on displacement and rotation (often called the "geometric" boundary conditions). It is not necessary that they satisfy the force and moment boundary conditions, but satisfaction of them may improve accuracy. However, it is not easy, in general, to find functions that satisfy all boundary conditions.
2. Each function must be continuous and p times differentiable, where p is the order of the highest spatial derivative in the Lagrangean. The pth derivative of at least one function must be nonzero. Here, from Eq. (3.297), $p = 2$.
3. If more than one function is used, they must be chosen from a set of functions that is complete. This means that any function on the interval $0 \le x \le \ell$ with the same boundary conditions as the problem under consideration can be expressed to any degree of accuracy as a linear combination of the functions in the set.

Examples of complete sets of functions on the interval $0 \leq x \leq \ell$ include

$$1, x, x^2, \ldots$$

$$\sin\left(\frac{\pi x}{\ell}\right), \sin\left(\frac{2\pi x}{\ell}\right), \sin\left(\frac{3\pi x}{\ell}\right), \ldots$$

a set of mode shapes for *any* problem

Completeness also implies that there can be no missing terms between the lowest and highest terms used in any series.

4. The set of functions must be linearly independent. This means that

$$\sum_{i=0}^{n} a_i \phi_i(x) = 0 \Rightarrow a_i = 0 \text{ for all } i \tag{3.300}$$

A set of functions that satisfies all of these criteria is said to be "admissible."

By use of the series approximation, we reduced a problem with an infinite number of degrees of freedom to one with n degrees of freedom. Instead of being governed by a partial differential equation, the behavior of this system is now defined by n second-order, ordinary differential equations in time. This reduction from a continuous system modeled by a partial differential equation with an infinite number of degrees of freedom to a system described by a finite number of ordinary differential equations in time is sometimes called spatial discretization. The number n is usually increased until convergence is obtained. (Note that if inertial forces are not considered so that the kinetic energy is identically zero, then a system described by an ordinary differential equation in a single spatial variable is reduced by the Ritz method to a system described by n algebraic equations.)

Now, let us illustrate how the approximating functions are actually used. Let ϕ_i, $i = 1, 2, \ldots, \infty$, be a complete set of p-times differentiable, linearly independent functions that satisfy the displacement and rotation boundary conditions. Thus, U can be written as

$$U = \frac{1}{2} \sum_{i=1}^{n} \sum_{j=1}^{n} \xi_i \xi_j \int_0^{\ell} \overline{EI} \phi_i'' \phi_j'' dx \tag{3.301}$$

The contributions of any springs that restrain the structure, as well as conservative loads, must be added to obtain the full potential energy P.

The kinetic energy of the beam is

$$K_{\text{beam}} = \frac{1}{2} \sum_{i=1}^{n} \sum_{j=1}^{n} \dot{\xi}_i \dot{\xi}_j \int_0^{\ell} m \phi_i \phi_j dx \tag{3.302}$$

Contributions of any additional particles and rigid bodies must be added to obtain the complete kinetic energy K.

The virtual work must account for distributed and concentrated forces resulting from all other sources, such as damping and aerodynamics. This can be written as

$$\overline{\delta W} = \sum_{i=1}^{n} \delta \xi_i \left[\int_0^\ell f(x,t)\phi_i \, dx + F_c(x_0,t)\phi_i(x_0) \right] \tag{3.303}$$

where x_0 is a value of x at which a concentrated force is located. Here, the first term accounts for a distributed force $f(x,t)$ on the interior of the beam, and the second term accounts for a concentrated force on the interior (see Eq. 3.96). In aeroelasticity, the loads $f(x,t)$ and $F_c(x_0,t)$ may depend on the displacement in a complicated manner.

The integrands in these quantities all involve the basis functions and their derivatives over the length of the beam. Note that these integrals involve only known quantities and often can be evaluated analytically. Sometimes they are too complicated to undertake analytically, however, and they must be evaluated numerically. Numerical evaluation is often facilitated by nondimensionalization. Symbolic computation tools such as Mathematica™ and Maple™ may be helpful in both situations.

With all such things considered, the equations of motion can be written in a form that is quite common; viz.

$$[M]\{\ddot{\xi}\} + [C]\{\dot{\xi}\} + [K]\{\xi\} = \{F\} \tag{3.304}$$

where $\{\xi\}$ is a column matrix of the generalized coordinates, $\{F\}$ is a column matrix of the generalized force terms that do not depend on ξ_i, $(\,\dot{}\,)$ is the time derivative of $(\,)$, $[M]$ is the mass matrix, $[C]$ is the gyroscopic/damping matrix, and $[K]$ is the stiffness matrix. The most important contribution to $[M]$ is from the kinetic energy, and this contribution is symmetric. The most important contribution to $[K]$ is from the strain energy of the structure and potential energy of any springs that restrain the motion of the structure. There can be contributions to all terms in the equations of motion from kinetic energy and virtual work. For example, there are contributions from kinetic energy to $[C]$ and $[K]$ when there is a rotating coordinate system. Damping makes contributions to $[C]$ through the virtual work. Finally, because aerodynamic loads, in general, depend on the displacement and its time derivatives, aeroelastic analyses may contain terms in $[M]$, $[C]$, and $[K]$ that stem from aerodynamic loads.

An interesting special case of this method occurs when the system is conservatively loaded. The resulting method is usually referred to as the Rayleigh–Ritz method, and many theorems can be proved about the convergence of approximations to the natural frequency. Indeed, one of the most powerful of such theorems states that the approximate natural frequencies are always upper bounds; another states that adding more terms to a given series always lowers the approximate natural frequencies (i.e., making them closer to the exact values).

A further specialized case is the simplest approximation, in which only one term is used. Then, an approximate expression for the lowest natural frequency can be written as a ratio called the "Rayleigh quotient." This simplest special case is of more than merely academic interest: It is not at all uncommon that a rough estimate of the lowest natural frequency is needed early in the design of flexible structures.

Example: The Ritz Method Using Clamped-Free Modes. In the first example, we consider a uniform, clamped-free beam that we modify by adding a tip mass of mass $\mu m\ell$. The exact solution can be obtained easily for this modified problem using the methodology described previously. However, it is desired here to illustrate the Ritz method, and we already calculated the modes for a clamped-free beam (i.e., without a tip mass) in Section 3.3.4. These mode shapes are solutions of an eigenvalue problem; therefore, provided we do not omit any modes between the lowest and highest mode number that we use, this set is automatically complete. The set is also orthogonal and therefore linearly independent. Of course, these modes automatically satisfy the boundary conditions on displacement and rotation for our modified problem (because they are the same as for the clamped-free beam), and they are infinitely differentiable. Hence, they are admissible functions for the modified problem. Moreover, they satisfy the condition of zero moment at the free end, which is a boundary condition for our modified problem. However, because of the presence of the tip mass in the modified problem, the shear force—which readers will recall is proportional to the third derivative of the displacement—does not vanish as it does for clamped-free mode shapes.

The strain energy becomes

$$U = \frac{1}{2} \sum_{i=1}^{n} \sum_{j=1}^{n} \xi_i \xi_j \int_0^\ell \overline{EI} \phi_i'' \phi_j'' dx \tag{3.305}$$

Substituting the mode shapes of Eq. (3.258) into Eq. (3.305) and taking advantage of orthogonality, we can simplify it to

$$U = \frac{\ell \overline{EI}}{2} \sum_{i=1}^{n} \xi_i^2 \alpha_i^4 \tag{3.306}$$

where α_i is the set of constants in Table 3.1. Similarly, accounting for the tip mass, the kinetic energy of which is

$$K_{\text{tip mass}} = \frac{1}{2} \mu m\ell \left[\frac{\partial v}{\partial t}(\ell, t) \right]^2$$

$$= \frac{1}{2} \mu m\ell \sum_{i=1}^{n} \sum_{j=1}^{n} \dot{\xi}_i \dot{\xi}_j \phi_i(\ell) \phi_j(\ell) \tag{3.307}$$

we obtain the total kinetic energy as

$$K = \frac{1}{2} \sum_{i=1}^{n} \sum_{j=1}^{n} \dot{\xi}_i \dot{\xi}_j \left[\int_0^\ell m \phi_i \phi_j dx + \mu m\ell \phi_i(\ell) \phi_j(\ell) \right] \tag{3.308}$$

With the use of the mode shapes in Eq. (3.258), we find that $\phi_i(\ell) = 2(-1)^{i+1}$; therefore, the kinetic energy simplifies to

$$K = \frac{m\ell}{2} \sum_{i=1}^{n} \sum_{j=1}^{n} \dot{\xi}_i \dot{\xi}_j \left[\delta_{ij} + 4\mu(-1)^{i+j} \right] \tag{3.309}$$

Table 3.3. *Approximate values of* $\omega_1\sqrt{\frac{m\ell^4}{EI}}$ *for*
clamped-free beam with tip mass of $\mu m\ell$ *using n*
clamped-free modes of Section 3.3.4, Eq. (3.258)

n	$\mu = 1$	$\mu = 10$	$\mu = 100$
1	1.57241	0.549109	0.175581
2	1.55964	0.542566	0.173398
3	1.55803	0.541748	0.173126
4	1.55761	0.541536	0.173055
5	1.55746	0.541458	0.173029
Exact	1.55730	0.541375	0.173001

where the Kronecker symbol $\delta_{ij} = 1$ for $i = j$ and $\delta_{ij} = 0$ for $i \neq j$. For free vibration, there are no additional forces. Thus, Lagrange's equations now can be written in matrix form as

$$[M]\{\ddot{\xi}\} + \lceil K_\cdot \rfloor \{\xi\} = 0 \tag{3.310}$$

where $\lceil K_\cdot \rfloor$ is a diagonal matrix with the diagonal elements given by

$$K_{ii} = \overline{EI}\ell\alpha_i^4 \qquad i = 1, 2, \ldots, n \tag{3.311}$$

and $[M]$ is a symmetric matrix with elements given by

$$M_{ij} = m\ell\left[\delta_{ij} + 4\mu(-1)^{i+j}\right] \qquad i, j = 1, 2, \ldots, n \tag{3.312}$$

Assuming $\xi = \overline{\xi}\exp(i\omega t)$, we can write Eq. (3.310) as an eigenvalue problem of the form

$$\left[[K] - \omega^2[M]\right]\{\overline{\xi}\} = 0 \tag{3.313}$$

Results for the first modal frequency are shown in Table 3.3 and compared therein with the exact solution. As we can see, the approximate solution agrees with the exact solution to within engineering accuracy with only two terms. For contrast, results for the second modal frequency are shown in Table 3.4; these results are not nearly as accurate. Results for the higher modes (not shown) are even less accurate. This is one of the problems with modal-approximation methods; fortunately, however,

Table 3.4. *Approximate values of* $\omega_2\sqrt{\frac{m\ell^4}{EI}}$ *for*
clamped-free beam with tip mass of $\mu m\ell$ *using n*
clamped-free modes of Section 3.3.4, Eq. (3.258)

n	$\mu = 1$	$\mu = 10$	$\mu = 100$
2	16.5580	15.8657	15.7867
3	16.3437	15.6191	15.5367
4	16.2902	15.5576	15.4744
5	16.2708	15.5353	15.4518
Exact	16.2501	15.5115	15.4277

Table 3.5. *Approximate values of* $\omega_1 \sqrt{\frac{m\ell^4}{EI}}$ *for clamped-free beam with tip mass of* $\mu m\ell$ *using n polynomial functions*

n	$\mu = 1$	$\mu = 10$	$\mu = 100$
1	1.55812	0.541379	0.173001
2	1.55733	0.541375	0.173001
3	1.55730	0.541375	0.173001
4	1.55730	0.541375	0.173001
5	1.55730	0.541375	0.173001
Exact	1.55730	0.541375	0.173001

aeroelasticians and structural dynamicists frequently are interested in only the lower-frequency modes. Note that the one-term approximation (i.e., the Rayleigh quotient) is within 1.1% for all values of μ displayed.

Example: The Ritz Method Using a Simple Power Series. As an alternative to using the mode shapes of a closely related problem, let us repeat the previous solution using a simple power series to construct a series of functions ϕ_i. Because the moment vanishes at the free end where $x = \ell$, we can make the second derivative of all terms proportional to $\ell - x$. To obtain a complete series, we can multiply this term by a complete power series $1, x, x^2$, and so on. Thus, we then may write the second derivative of the ith function as

$$\phi_i'' = \frac{1}{\ell^2} \left(1 - \frac{x}{\ell}\right) \left(\frac{x}{\ell}\right)^{i-1} \tag{3.314}$$

With the boundary conditions on displacement and rotation being $\phi_i(0) = \phi_i'(0) = 0$, we then can integrate to find an expression for the ith function as

$$\phi_i = \frac{\left(\frac{x}{\ell}\right)^{i+1} \left[2 + i - i\left(\frac{x}{\ell}\right)\right]}{i\,(1+i)\,(2+i)} \tag{3.315}$$

Because the chosen admissible functions have nonzero third derivatives at the tip, they offer the possibility of satisfying the nonzero shear condition in combination with one another. Such admissible functions are sometimes called "quasi-comparison functions."

In this case, the stiffness matrix becomes

$$K_{ij} = \frac{2\overline{EI}}{\ell^3 (i + j - 1)(i + j)(1 + i + j)} \qquad i, j = 1, 2, \ldots, n \tag{3.316}$$

and the mass matrix

$$M_{ij} = \frac{2m\ell \left[3(i^2 + j^2) + 7ij + 23(i + j) + 40\right]}{ij\,(i+1)\,(i+2)\,(j+1)\,(j+2)\,(i+j+3)\,(i+j+4)\,(i+j+5)}$$
$$+ \frac{4\,\mu m\ell}{ij\,(i+1)\,(i+2)\,(j+1)\,(j+2)} \qquad i, j = 1, 2, \ldots, n \tag{3.317}$$

Results from this calculation are given in Tables 3.5 and 3.6 for the first two modes. It is clear that these results are *much* better than those obtained with the

Table 3.6. *Approximate values of* $\omega_2 \sqrt{\frac{m\ell^4}{EI}}$ *for clamped-free beam with tip mass of* $\mu m \ell$ *using n polynomial functions*

n	$\mu = 1$	$\mu = 10$	$\mu = 100$
2	16.2853	15.5443	15.4605
3	16.2841	15.5371	15.4524
4	16.2505	15.5119	15.4280
5	16.2501	15.5116	15.4277
Exact	16.2501	15.5115	15.4277

clamped-free beam modes. It is not unusual for polynomial functions to provide better results than those obtained with beam mode shapes. However, here it is worth noting that the beam mode shapes are at a disadvantage for this problem. Unlike the problem being solved (and the polynomials chosen), the beam mode shapes are constrained to have zero shear force at the free end and thus are not quasi-comparison functions for the problem with a tip mass. This one-term polynomial approximation (i.e., the Rayleigh quotient) is within 0.05%, which is exceptionally good given its simplicity.

It is sometimes suggested that the mode shapes of a closely related problem are—at least, in some sense—superior to other approximate sets of functions. For example, in the first example, we saw that the orthogonality of the modes used resulted in a diagonal stiffness matrix, which provides a slight advantage in the ease of computing the eigenvalues. However, for the low-order problems of the sort we are discussing, that advantage is hardly noticeable. Indeed, symbolic computation tools such as Mathematica™ and Maple™ are capable of calculating the eigenvalues for problems of the size of this example in but a few seconds. Moreover, in some cases, the simplicity of carrying out the integrals that result in approximate formulations is a more important factor in deciding which set of functions to use in a standard implementation of the Ritz method. Indeed, polynomial functions are generally much easier to deal with analytically than free-vibration modes such as those illustrated in Section 3.5.1, which frequently involve transcendental functions.

Alternatives to the standard Ritz method include the methods of Galerkin, finite elements, component mode synthesis, flexibility influence coefficients, methods of weighted residuals, collocation methods, and integral equation methods. We introduce Galerkin's method and the finite element method in the next two sections. Detailed descriptions of other approaches are found in more advanced texts on structural dynamics and aeroelasticity.

3.5.2 Galerkin's Method

Rather than making use of energy and Lagrange's equation as in the Ritz method, Galerkin's method starts with the partial differential equation of motion. Let us denote this equation by

$$\mathcal{L}[v(x, t)] = 0 \tag{3.318}$$

where $\mathcal{L}$ is an operator on the unknown function $v(x, t)$ with maximum spatial partial derivatives of the order q. For the structural dynamics problems addressed so far, the operator $\mathcal{L}$ is linear and $q = 2p$, where p is the maximum order of spatial partial derivative in the Lagrangean. It is important to note, however, that it is not true, in general, that $q = 2p$; indeed, we do not need to consider the Lagrangean at all with this method.

To apply Galerkin's method, we need to express $v(x, t)$ and, hence, the operator $\mathcal{L}$ in terms of a series of functions with one or more terms. For a beam in bending, for example, this means that, as before

$$v(x, t) = \sum_{j=1}^{n} \xi_j(t)\phi_j(x) \tag{3.319}$$

Relative to the basis functions used in the Ritz method, the characteristics that these functions ϕ_i must possess for use in Galerkin's method are more stringent, as follows:

1. Each function must satisfy all boundary conditions. Note that it is not easy, in general, to find functions that satisfy all boundary conditions.
2. Each function must be at least q times differentiable. The qth derivative of at least one function must be nonzero.
3. If more than one function is used, they must be chosen from a set of functions that is complete.
4. The set of functions must be linearly independent.

Functions that satisfy all of these criteria are said to be "comparison functions." The original partial differential equation then is multiplied by ϕ_i and integrated over the domain of the independent variable (e.g., $0 \le x \le \ell$). Thus, a set of n ordinary differential equations is obtained from the original partial differential equation. (Note that if the original equation is an ordinary differential equation in x, then Galerkin's method yields n algebraic equations.)

Consider a beam in bending as an example. The equation of motion can be written as in Eq. (3.194), with a slight change, as

$$\frac{\partial^2}{\partial x^2}\left(\overline{EI}\frac{\partial^2 v}{\partial x^2}\right) + m\frac{\partial^2 v}{\partial t^2} - f(x, t) = 0 \tag{3.320}$$

where $\overline{EI}$ is the flexural rigidity, m is the mass per unit length, and the boundary conditions and loading term $f(x, t)$ must reflect any attached particles or rigid bodies. In aeroelasticity, the loads $f(x, t)$ may depend on the displacement in a complicated manner.

With all of the components as described herein considered, the discretized equations of motion can be written in the same form as in the Ritz method; that is

$$[M]\{\ddot{\xi}\} + [C]\{\dot{\xi}\} + [K]\{\xi\} = \{F\} \tag{3.321}$$

where $\{\xi\}$ is a column matrix of the generalized coordinates, $\{F\}$ is a column matrix of the generalized force terms that do not depend on ξ_i, $(\dot{\ })$ is the time derivative of $(\ )$, $[M]$ is the mass matrix, $[C]$ is the gyroscopic/damping matrix, and $[K]$ is the stiffness matrix. As before, inertial forces contribute to $[M]$, there are contributions from the inertial forces to $[C]$ and $[K]$ when there is a rotating coordinate system, and damping also contributes to $[C]$. Finally, because aeroelastic loads, in general, depend on the displacement and its time derivatives, aerodynamics can contribute terms to $[M]$, $[C]$, and $[K]$.

Example: Galerkin's Method for a Beam in Bending. Now, we illustrate how the approximating functions are actually used. Let ϕ_i, $i = 1, 2, \ldots, \infty$ be a complete set of q-times differentiable, linearly independent functions that satisfy all of the boundary conditions. Substituting Eq. (3.319) into Eq. (3.320), multiplying by $\phi_i(x)$, and integrating over x from 0 to ℓ, we obtain

$$\int_0^\ell \phi_i \left[\sum_{j=1}^n \xi_j (\overline{EI}\phi_j'')'' + \sum_{j=1}^n \ddot{\xi}_j m\phi_j - f(x, t) \right] dx = 0 \quad i = 1, 2, \ldots, n \quad (3.322)$$

After reversing the order of integration and summation and integrating the first term by parts, and taking into account that the functions ϕ_i satisfy all the boundary conditions, this equation becomes

$$\sum_{j=1}^n \left(\xi_j \int_0^\ell \overline{EI}\phi_i''\phi_j'' dx + \ddot{\xi}_j \int_0^\ell m\phi_i\phi_j dx \right) - \int_0^\ell f\phi_i dx = 0 \quad i = 1, 2, \ldots, n$$

$$(3.323)$$

When we compare the first two terms with the previous derivation by the Ritz method, we see the close relationship between these approaches. Indeed, if the starting partial differential equation is derivable from energy—which implies that $q = 2p$—and the same approximating functions ϕ_i are used in both cases, the resulting discretized equations are the same.

Considering the clamped-free case, for example, we can develop a set of comparison functions by starting with

$$\phi_i'' = \frac{1}{\ell^2} \left(1 - \frac{x}{\ell}\right)^2 \left(\frac{x}{\ell}\right)^{i-1} \quad (3.324)$$

With the boundary conditions on displacement and rotation being $\phi_i(0) = \phi_i'(0) = 0$, we then can integrate to find an expression for the ith function as

$$\phi_i = \frac{\left(\frac{x}{\ell}\right)^{1+i} \left\{6 + i^2 \left(1 - \frac{x}{\ell}\right)^2 + i\left[5 - \frac{6x}{\ell} + \left(\frac{x}{\ell}\right)^2\right]\right\}}{i(1+i)(2+i)(3+i)} \quad (3.325)$$

Elements of the stiffness matrix are found as

$$K_{ij} = \int_0^\ell \overline{EI}\phi_i''\phi_j'' dx$$

$$= \frac{24\overline{EI}}{\ell^3 (i+j-1)(i+j)(1+i+j)(2+i+j)(3+i+j)} \quad (3.326)$$

Table 3.7. *Approximate values of* $\omega_i\sqrt{\frac{m\ell^4}{EI}}$ *for*
$i = 1, 2,$ *and* 3, *for a clamped-free beam using*
n polynomial functions

n	Mode 1	Mode 2	Mode 3
1	3.53009	–	–
2	3.51604	22.7125	–
3	3.51602	22.0354	66.2562
4	3.51602	22.0354	61.7675
5	3.51602	22.0345	61.7395
exact	3.51602	22.0345	61.6972

Similarly, the elements of the mass matrix are found as

$$M_{ij} = \int_0^\ell m\phi_i\phi_j dx$$

$$= \frac{m\ell p_1}{p_2} \tag{3.327}$$

where

$$p_1 = 30,240 + 28,512(i + j) + 9,672(i^2 + j^2) + 1,392(i^3 + j^3) + 72(i^4 + j^4)$$

$$+ 20,040ij + 4,520(i^2 j + ij^2) + 320(i^3 j + ij^3) + 520i^2 j^2$$

$$p_2 = i(1 + i)(2 + i)(3 + i)j(1 + j)(2 + j)(3 + j)(3 + i + j)$$

$$(4 + i + j)(5 + i + j)(6 + i + j)(7 + i + j) \tag{3.328}$$

The fact that the governing equation is derivable from energy is reflected in the symmetry of $[M]$ and $[K]$. Results for free vibration (i.e., with $f = 0$) are given in Table 3.7. As with the Ritz method, we see monotonic convergence from above and accuracy comparable to that achieved via the Ritz method. However, unlike the Ritz method, we do not always obtain results for free-vibration problems that converge from above.

Example: Galerkin's Method for a Beam in Bending Using an Alternative Form of the Equation of Motion. Consider again a clamped-free beam. To obtain an alternative equation of motion, we integrate the equation of motion twice and use the boundary conditions of zero shear and bending moment to obtain an integro-partial differential equation

$$\overline{EI}\frac{\partial^2 v}{\partial x^2} + \int_x^\ell (x - \zeta)\left[f(\zeta, t) - m\frac{\partial^2 v(\zeta, t)}{\partial t^2} \right] d\zeta = 0 \tag{3.329}$$

where ζ is a dummy variable. Although this equation of motion is somewhat more complicated, it is only a second-order equation. Thus, it has only two boundary conditions, which are zero displacement and slope at $x = 0$. Thus, a much simpler

Table 3.8. *Approximate values of $\omega_i \sqrt{\frac{m\ell^4}{EI}}$ for $i = 1, 2,$ and 3 for a clamped-free beam using n terms of a power series with a reduced-order equation of motion*

n	Mode 1	Mode 2	Mode 3
1	7.48331	–	–
2	3.84000	57.2822	–
3	3.44050	24.1786	188.677
4	3.52131	20.3280	69.3819
5	3.51698	22.0793	53.2558
6	3.51607	22.1525	61.0295
exact	3.51602	22.0345	61.6972

set of comparison functions can be used, such as a simple power series; that is

$$\phi_i = \left(\frac{x}{\ell}\right)^{i+1} \quad i = 1, 2, \ldots, n \tag{3.330}$$

We should not expect greater accuracy from this simple set of functions, but the analytical effort is considerably less. Indeed, the elements of the stiffness matrix are

$$K_{ij} = \int_0^\ell \overline{EI}\phi_i\phi_j'' dx$$
$$= \frac{\overline{EI}j(j+1)}{\ell(i+j+1)} \tag{3.331}$$

and the elements of the mass matrix are

$$M_{ij} = \int_0^\ell \phi_i \int_x^\ell (\zeta - x)m\phi_j(\zeta)d\zeta \, dx$$
$$= \frac{m\ell^3}{(2+i)(3+i)(5+i+j)} \tag{3.332}$$

Note that these matrices are not symmetric. Moreover, the results presented in Table 3.8 are not as accurate as those obtained in Table 3.7, and the convergence is not monotonic from above.

The partial differential equations derived previously for free vibration of strings, beams in torsion, and beams in bending can be derived from energy-based approaches, such as Hamilton's principle. (The use of Hamilton's principle is beyond the scope of this text, but detailed treatments are found in numerous graduate-level texts on structural dynamics.) In those cases, the Ritz and Galerkin's methods give the same results when used with the same approximating functions. As shown here, however, Galerkin's method provides a viable alternative to the Ritz method in cases where the equations of motion are not of the form presented previously in this chapter.

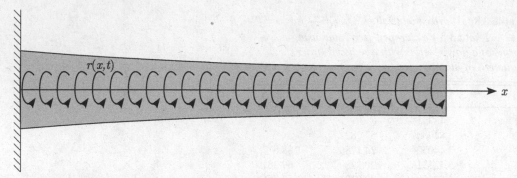

Figure 3.44. Schematic of a nonuniform beam with distributed twisting moment per unit length

3.5.3 The Finite Element Method

The finite element method is, by far, the most popular way of solving realistic structural dynamics and aeroelasticity problems in industry. The name derives from the breaking of a structure into a large number of small elements, modeling them approximately, and connecting them together appropriately. Because of this way of discretizing the geometry, it is possible to accurately capture modeling details that other methods cannot.

In one sense, the finite element method can be regarded as a special case of Ritz and Galerkin methods, one in which the generalized coordinates are themselves displacements and/or rotations at points along the structure. It typically makes use of polynomial shape functions over each of the finite elements into which the original structure is broken. Equations based on the finite element method have the same structure as Eq. (3.304); however, they are typically of large order, with n being on the order of 10^2 to 10^7. What keeps the computational effort from being overly burdensome is that the matrices have a narrow-banded structure, which allows specialized software to be used in solving the equations of motion that takes advantage of this structure, reducing both memory and floating-point operations and resulting in significant computational advantages.

Here, we present only a simple outline of the method as applied to beams in torsion and in bending, leaving more advanced topics such as plates and shells to textbooks devoted to the finite element method, such as those by Reddy (1993) and Zienkiewicz and Taylor (2005).

Application to Beams in Torsion. Here, we use the finite element method to analyze the behavior of a nonuniform beam in torsion. Similar to the application of the Ritz method, we make use of Lagrange's equation. Regardless of how finite elements are derived, however, for a sufficiently fine mesh, the results should approach the exact structural behavior. This development encompasses both forced response and free vibration.

Consider a clamped-free beam subjected to a distributed torque $r(x, t)$ as depicted in Fig. 3.44. Note that the x coordinate is along the beam. The strain energy

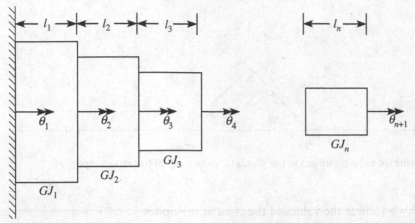

Figure 3.45. Schematic of a nonuniform beam with internal torques discretized

of the system can be written as

$$U = \frac{1}{2} \int_0^{\ell} \overline{GJ}(x) \left(\frac{\partial \theta}{\partial x} \right)^2 dx \qquad (3.333)$$

where $\overline{GJ}(x)$ is the torsional stiffness of the wing and $\theta(x, t)$ is the elastic twist. In the finite-element approach, the beam is divided into n elements, as shown in Fig. 3.45. Although there is no requirement to make the elements of constant stiffness, we do so for convenience. Relaxation of this assumption is left as an exercise for readers (see Problem 25). Element i is connected to two end nodes i and $i + 1$ with coordinates x_i and x_{i+1}, respectively. Within element i, the torsional stiffness is assumed to be a constant, $\overline{GJ}_i$. The discrete value of the twist at the node i is denoted θ_i. The twist is linearly interpolated between the nodal values so that

$$\theta(x, t) = \begin{Bmatrix} 1 - z \\ z \end{Bmatrix}^T \begin{Bmatrix} \theta_i(t) \\ \theta_{i+1}(t) \end{Bmatrix} \qquad (3.334)$$

where

$$z = \frac{x - x_i}{\ell_i} \qquad (3.335)$$

with $0 \leq z \leq 1$. The expression for $\theta(x, t)$ also can be written as

$$\theta(x, t) = \theta_i(t) + \frac{(x - x_i)}{\ell_i} [\theta_{i+1}(t) - \theta_i(t)] \qquad (3.336)$$

where $x_i \leq x \leq x_{i+1}$ and $\ell_i = x_{i+1} - x_i$. Note that if all θ_i are zero except one, then only the element immediately to the left (element $i - 1$) and immediately to the right (element i) are affected (Fig. 3.46). Introducing this approximation into the strain energy, Eq. (3.333), and integrating over the beam length yields

$$U = \frac{1}{2} \{\theta\}^T [K] \{\theta\} \qquad (3.337)$$

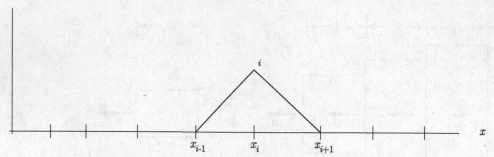

Figure 3.46. Assumed twist distribution for all nodal values equal to zero except θ_i

where the array $\{\theta\}$ stores the values of the twist at the nodes

$$\{\theta(t)\}^T = \lfloor \theta_1(t) \; \theta_2(t) \cdots \theta_{n+1}(t) \rfloor \tag{3.338}$$

The resulting stiffness matrix $[K]$ may be written as

$$[K] = \begin{bmatrix} \frac{\overline{GJ}_1}{\ell_1} & -\frac{\overline{GJ}_1}{\ell_1} & 0 & 0 & 0 & 0 & \cdots \\ -\frac{\overline{GJ}_1}{\ell_1} & \frac{\overline{GJ}_1}{\ell_1} + \frac{\overline{GJ}_2}{\ell_2} & -\frac{\overline{GJ}_2}{\ell_2} & 0 & 0 & 0 & \cdots \\ 0 & -\frac{\overline{GJ}_2}{\ell_2} & \frac{\overline{GJ}_2}{\ell_2} + \frac{\overline{GJ}_3}{\ell_3} & -\frac{\overline{GJ}_3}{\ell_3} & 0 & 0 & \cdots \\ 0 & 0 & -\frac{\overline{GJ}_3}{\ell_3} & \frac{\overline{GJ}_3}{\ell_3} + \frac{\overline{GJ}_4}{\ell_4} & -\frac{\overline{GJ}_4}{\ell_4} & \ddots & \ddots \\ 0 & 0 & 0 & -\frac{\overline{GJ}_4}{\ell_4} & \frac{\overline{GJ}_4}{\ell_4} + \frac{\overline{GJ}_5}{\ell_5} & \ddots & \ddots \\ 0 & 0 & 0 & \ddots & & \ddots & \ddots \\ \vdots & \vdots & \vdots & \ddots & \ddots & \ddots & \ddots \end{bmatrix} \tag{3.339}$$

Note that we could add the potential energy of springs attached to ground at any nodes to represent elastic restraints.

The kinetic energy may be written as

$$K = \frac{1}{2} \int_0^\ell \overline{\rho I_p}(x) \left(\frac{\partial \theta}{\partial t} \right)^2 dx \tag{3.340}$$

Using the same interpolation for $\theta(x, t)$ and a constant mass polar moment of inertia per unit length in $\overline{\rho I_{p_i}}$ in element i, we obtain a discretized kinetic energy of the form

$$K = \frac{1}{2} \{\dot{\theta}\}^T [M] \{\dot{\theta}\} \tag{3.341}$$

where the mass matrix $[M]$ is given by

$$[M] = \begin{bmatrix} \frac{\rho I_{p_1} \ell_1}{3} & \frac{\rho I_{p_1} \ell_1}{6} & 0 & 0 & 0 & 0 & \cdots \\ \frac{\rho I_{p_1} \ell_1}{6} & \frac{\rho I_{p_1} \ell_1}{3} + \frac{\rho I_{p_2} \ell_2}{3} & \frac{\rho I_{p_2} \ell_2}{6} & 0 & 0 & 0 & \cdots \\ 0 & \frac{\rho I_{p_2} \ell_2}{6} & \frac{\rho I_{p_2} \ell_2}{3} + \frac{\rho I_{p_3} \ell_3}{3} & \frac{\rho I_{p_3} \ell_3}{6} & 0 & 0 & \cdots \\ 0 & 0 & \frac{\rho I_{p_3} \ell_3}{6} & \frac{\rho I_{p_3} \ell_3}{3} + \frac{\rho I_{p_4} \ell_4}{3} & \frac{\rho I_{p_4} \ell_4}{6} & \ddots & \ddots \\ 0 & 0 & 0 & \frac{\rho I_{p_4} \ell_4}{6} & \frac{\rho I_{p_4} \ell_4}{3} + \frac{\rho I_{p_5} \ell_5}{3} & \ddots & \ddots \\ 0 & 0 & 0 & \ddots & \ddots & \ddots & \ddots \\ \vdots & \vdots & \vdots & \ddots & \ddots & \ddots & \ddots \end{bmatrix} \tag{3.342}$$

We also could add concentrated inertia at any nodes to represent the inertia of any attached rigid bodies.

The contribution of the applied torque $r(x, t)$ comes into the analysis through the generalized force, which may be extracted from the virtual work, given by Eq. (2.46) and repeated here for convenience as

$$\overline{\delta W} = \int_0^\ell r(x, t) \delta \theta(x, t) dx \tag{3.343}$$

Here, it is helpful to represent the twisting moment using the same shape functions as for θ, viz.

$$r(x, t) = r_i(t) + \frac{(x - x_i)}{\ell_i} [r_{i+1}(t) - r_i(t)] \tag{3.344}$$

with the array

$$\{r\}^T = \lfloor r_1 \, r_2 \cdots r_{n+1} \rfloor \tag{3.345}$$

representing the nodal values of the applied torque per unit span. The virtual work then becomes

$$\overline{\delta W} = \{\delta \theta\}^T [D] \{r(t)\} \tag{3.346}$$

so that the generalized force then may be put into the form

$$\{\Xi\} = [D] \{r(t)\} \tag{3.347}$$

with the loading matrix $[D]$ given by

$$[D] = \begin{bmatrix} \frac{\ell_1}{3} & \frac{\ell_1}{6} & 0 & 0 & 0 & 0 & \cdots \\ \frac{\ell_1}{6} & \frac{\ell_1}{3}+\frac{\ell_2}{3} & \frac{\ell_2}{6} & 0 & 0 & 0 & \cdots \\ 0 & \frac{\ell_2}{6} & \frac{\ell_2}{3}+\frac{\ell_3}{3} & \frac{\ell_3}{6} & 0 & 0 & \cdots \\ 0 & 0 & \frac{\ell_3}{6} & \frac{\ell_3}{3}+\frac{\ell_4}{3} & \frac{\ell_4}{6} & \ddots & \ddots \\ 0 & 0 & 0 & \frac{\ell_4}{6} & \frac{\ell_4}{3}+\frac{\ell_5}{3} & \ddots & \ddots \\ 0 & 0 & 0 & \ddots & & \ddots & \ddots \\ \vdots & \vdots & \vdots & \ddots & \ddots & \ddots & \ddots \end{bmatrix} \qquad (3.348)$$

As for the boundary conditions, admissibility requires only that we satisfy the geometric boundary conditions (see Section 3.5.1). If we consider, for example, a clamped-free beam, we need only set $\theta_1 = 0$. The boundary condition at the free end (i.e., zero twisting moment) is a "natural" (i.e., a force or moment) boundary condition, not a geometric condition. Therefore, it need not be taken into account in a solution by this approach. As a consequence, the first elements of column matrices $\{\theta\}$ and $\{r\}$ are removed; the reason the first element of the latter is removed is that $\delta\theta_1 = 0$, in keeping with the requirement that the virtual displacements and rotations must satisfy the geometric boundary conditions. This has the effect of removing the first row and column from each of the matrices $[M]$ and $[K]$, and the first row from $[D]$.

The equations of motion now may be formed by use of Lagrange's equation, as with the Ritz method. Given the approximation of the twist field in Eq. (3.336), the only unknowns of the problem are the nodal twist angles θ_i. Thus, the equations of motion may be written as:

$$\frac{d}{dt}\left(\frac{\partial K}{\partial\{\dot\theta\}}\right) + \frac{\partial U}{\partial\{\theta\}} = \{\Xi\} \qquad (3.349)$$

or

$$[M]\{\ddot\theta\} + [K]\{\theta\} = [D]\{r(t)\} \qquad (3.350)$$

Although the size of the system matrices in the finite element method can be very large, these matrices possess important properties. First, as noted previously in the discussion of the Ritz method (see Section 3.5.1), they are symmetric, which here is a reflection of their having been derived from energy methods applied to conservative systems. Second, they are banded; that is, the nonvanishing entries are concentrated around the diagonals of the matrices. Third, $[M]$ is positive definite and $[K]$ is at least positive semidefinite. In the absence of rigid-body modes, $[K]$ is positive definite because it results from the computation of the strain energy of the structure, itself a positive-definite quantity when rigid-body motion is excluded.

With these equations, we may look at several types of problems for the nonuniform beam in torsion. For example:

1. The static response of the beam may be found if $\{r\}$ is not a function of time. For this, we do not need the mass matrix $[M]$. Thus

$$[K]\{\theta\} = [D]\{r\} \tag{3.351}$$

2. The free-vibration characteristics of the beam may be found by setting $r = 0$, assuming simple harmonic motion such that $\{\theta\} = \{\hat{\theta}\}\exp(i\omega t)$ and solving the eigenvalue problem

$$[K]\{\hat{\theta}\} = \omega^2[M]\{\hat{\theta}\} \tag{3.352}$$

3. If $\{r(t)\}$ has the form $\{\hat{r}\}\exp(i\Omega t)$ with $\{\hat{r}\}$ and Ω specified constants, the steady-state response to harmonic excitation may be found by assuming $\{\theta\}(t) = \{\hat{\theta}\}\exp(i\Omega t)$ and solving the algebraic equations

$$[[K] - \Omega^2[M]]\{\hat{\theta}\} = [D]\{\hat{r}\} \tag{3.353}$$

4. Finally, the forced response of the structure may be determined by numerical integration of Eqs. (3.350) subject to appropriate initial conditions—that is, specified values for $\theta_i(0)$ and $\dot{\theta}_i(0)$.

Complex structures including entire aircraft can be modeled with the finite element method. The resulting discretized equations are similar to Eq. (3.350), where $\{\theta\}$ is an array of nodal displacements and/or rotations, $\{r(t)\}$ an array of nodal forces and/or torques, $[K]$ is a stiffness matrix characterizing the elastic behavior of the entire structure, and $[M]$ is a mass matrix characterizing the inertia properties of the entire structure. As the complexity of the model increases, the various arrays increase in size. For the most general types of models, such as those based on three-dimensional brick elements, hundreds of thousands of degrees of freedom or more may be required to accurately model a complete wing structure.

As an illustrative example, results obtained for the tip rotation caused by twisting of a beam with linearly varying $\overline{GJ}(x)$ with $\overline{GJ}(0) = \overline{GJ}_0 = 2\overline{GJ}(\ell)$, $r(x, t) = r = \text{const.}$, and constant values of $\overline{GJ}$ within each element are presented in Table 3.9. The convergence is monotonic, and the answers are evidently upper bounds.

Application to Beams in Bending. As another example of applying the finite element method, we next turn to its application to beams in bending. The theory of bending for beams was presented in terms of strain energy, kinetic energy, and virtual work; this framework is sufficient for constructing a finite-element model for nonuniform beams in bending. Again, strictly for simplicity, we assume the bending stiffness and mass per unit length to be constants, respectively equal to $\overline{EI}_i$ and m_i within element i. Allowing for linearly varying $\overline{EI}$ and m within each element is

Table 3.9. *Finite-element results for the tip rotation caused by twist of a beam with linearly varying* $\overline{GJ}(x)$ *such that* $\overline{GJ}(0) = \overline{GJ}_0 = 2\overline{GJ}(\ell)$, $r(x,t) = r = const.$, *and constant values of* $\overline{GJ}$ *within each element*

n	$\theta(\ell)\dfrac{r\ell^2}{\overline{GJ}_0}$
1	0.666667
2	0.628571
3	0.620491
4	0.617560
5	0.616184
6	0.615431
7	0.614976
exact	0.613706

left as an exercise for readers (see Problem 26). We consider a beam loaded with a distributed force per unit length $f(x,t)$ and a distributed bending moment per unit length $q(x,t)$ as shown in Fig. 3.47.

As with the beam in torsion, we now develop the stiffness matrix from the strain energy. The strain energy is given by

$$U = \frac{1}{2} \int_0^\ell \overline{EI} \left(\frac{\partial^2 v}{\partial x^2} \right)^2 dx \tag{3.354}$$

where $v(x,t)$ is represented in terms of nodal displacements $v_i(t)$ and rotations $\beta_i(t)$. The latter is in the sense of the bending slope $\beta(x,t) = \partial v(x,t)/\partial x$. When x is between nodes i and $i+1$, $v(x,t)$ is approximated as

$$v(x,t) = \begin{Bmatrix} 2z^3 - 3z^2 + 1 \\ z^3 - 2z^2 + z \\ 3z^2 - 2z^3 \\ z^3 - z^2 \end{Bmatrix}^T \begin{Bmatrix} v_i(t) \\ \ell_i \beta_i(t) \\ v_{i+1}(t) \\ \ell_i \beta_{i+1}(t) \end{Bmatrix} \tag{3.355}$$

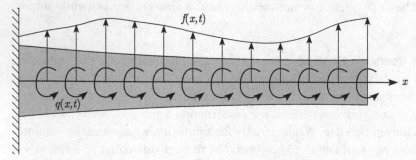

Figure 3.47. Schematic of a nonuniform beam with distributed force and bending moment per unit length

where

$$z = \frac{x - x_i}{\ell_i} \tag{3.356}$$

with $0 \le z \le 1$. The four cubic polynomials in Eq. (3.355) are called "Hermite polynomials." They have the property that one of their values or derivatives at the ends (i.e., where $z = 0$ or $z = 1$) is equal to unity whereas the other three are equal to zero. This way, the element degrees of freedom are displacements or rotations at the ends of the element. With this interpolation of $v(x, t)$, the strain energy can be written as

$$U = \frac{1}{2} \{\xi\}^T [K] \{\xi\} \tag{3.357}$$

where the degrees of freedom are arranged in the column matrix $\{\xi(t)\}$ of length $2n + 2$ so that

$$\{\xi(t)\} = \begin{Bmatrix} v_1(t) \\ \beta_1(t) \\ v_2(t) \\ \beta_2(t) \\ \vdots \\ v_{n+1}(t) \\ \beta_{n+1}(t) \end{Bmatrix} \tag{3.358}$$

and the stiffness matrix $[K]$ has the form

$$[K] = \begin{bmatrix} \frac{12EI_1}{\ell_1^3} & \frac{6EI_1}{\ell_1^2} & -\frac{12EI_1}{\ell_1^3} & \frac{6EI_1}{\ell_1^2} & 0 & 0 \\ \frac{6EI_1}{\ell_1^2} & \frac{4EI_1}{\ell_1} & -\frac{6EI_1}{\ell_1^2} & \frac{2EI_1}{\ell_1} & 0 & 0 \\ -\frac{12EI_1}{\ell_1^3} & -\frac{6EI_1}{\ell_1^2} & \frac{12EI_1}{\ell_1^3} + \frac{12EI_2}{\ell_2^3} & \frac{6EI_2}{\ell_2^2} - \frac{6EI_1}{\ell_1^2} & -\frac{12EI_2}{\ell_2^3} & \frac{6EI_2}{\ell_2^2} \\ \frac{6EI_1}{\ell_1^2} & \frac{2EI_1}{\ell_1} & \frac{6EI_2}{\ell_2^2} - \frac{6EI_1}{\ell_1^2} & \frac{4EI_1}{\ell_1} + \frac{4EI_2}{\ell_2} & -\frac{6EI_2}{\ell_2^2} & \frac{2EI_2}{\ell_2} \\ 0 & 0 & -\frac{12EI_2}{\ell_2^3} & -\frac{6EI_2}{\ell_2^2} & \frac{12EI_2}{\ell_2^3} & -\frac{6EI_2}{\ell_2^2} \\ 0 & 0 & \frac{6EI_2}{\ell_2^2} & \frac{2EI_2}{\ell_2} & -\frac{6EI_2}{\ell_2^2} & \frac{4EI_2}{\ell_2} \end{bmatrix} \tag{3.359}$$

for the two-element case. Note that the contributions from element 1 are all in the upper-left 4×4 submatrix, whereas the contributions from element 2 are all in the lower-right 4×4 submatrix. The two overlap at the 2×2 submatrix in the middle for degrees of freedom associated with the node at the right end of element 1 and at the left end of element 2. With this pattern in mind, it is a straightforward matter to expand the matrix to an arbitrary number of elements.

The kinetic energy is given by

$$K = \frac{1}{2} \int_0^\ell m \left(\frac{\partial v}{\partial t} \right)^2 dx \tag{3.360}$$

that with the specified interpolation can be written in discretized form as

$$K = \frac{1}{2}\{\dot{\xi}\}^T [M]\{\dot{\xi}\} \tag{3.361}$$

with the mass matrix given by

$$[M] = \begin{bmatrix} \frac{13}{35}\ell_1 m_1 & \frac{11}{210}\ell_1^2 m_1 & \frac{9}{70}\ell_1 m_1 & -\frac{13}{420}\ell_1^2 m_1 & 0 & 0 \\[6pt] \frac{11}{210}\ell_1^2 m_1 & \frac{1}{105}\ell_1^3 m_1 & \frac{13}{420}\ell_1^2 m_1 & -\frac{1}{140}\ell_1^3 m_1 & 0 & 0 \\[6pt] \frac{9}{70}\ell_1 m_1 & \frac{13}{420}\ell_1^2 m_1 & \frac{13}{35}(\ell_1 m_1 + l_2 m_2) & -\frac{11}{210}(l_1^2 m_1 - l_2^2 m_2) & \frac{9}{70}\ell_2 m_2 & -\frac{13}{420}\ell_2^2 m_2 \\[6pt] -\frac{13}{420}\ell_1^2 m_1 & -\frac{1}{140}\ell_1^3 m_1 & -\frac{11}{210}(l_1^2 m_1 - l_2^2 m_2) & \frac{1}{105}(m_1 \ell_1^3 + l_2^3 m_2) & \frac{13}{420}\ell_2^2 m_2 & -\frac{1}{140}\ell_2^3 m_2 \\[6pt] 0 & 0 & \frac{9}{70}\ell_2 m_2 & \frac{13}{420}\ell_2^2 m_2\ . & \frac{13}{35}\ell_2 m_2 & -\frac{11}{210}\ell_2^2 m_2 \\[6pt] 0 & 0 & -\frac{13}{420}\ell_2^2 m_2 & -\frac{1}{140}\ell_2^3 m_2 & -\frac{11}{210}\ell_2^2 m_2 & \frac{1}{105}\ell_2^3 m_2 \end{bmatrix} \tag{3.362}$$

again for the two-element case. This pattern is the same as that of the stiffness matrix, so it is also a straightforward matter to expand the mass matrix to an arbitrary number of elements.

Finally, the contributions of the applied distributed force and bending moment are determined using the virtual work. If we interpolate both $f(x, t)$ and $q(x, t)$ in the same way that $r(x, t)$ was treated for torsion, viz.

$$f(x, t) = f_i(t) + \frac{(x - x_i)}{\ell_i}[f_{i+1}(t) - f_i(t)]$$
$$q(x, t) = q_i(t) + \frac{(x - x_i)}{\ell_i}[q_{i+1}(t) - q_i(t)] \tag{3.363}$$

with the arrays $\{f(t)\}$ and $\{q(t)\}$ representing the nodal values of the applied force and bending moment per unit span

$$\{f\}^T = \lfloor f_1\ f_2 \cdots f_{n+1}\rfloor$$
$$\{q\}^T = \lfloor q_1\ q_2 \cdots q_{n+1}\rfloor \tag{3.364}$$

the virtual work then becomes

$$\overline{\delta W} = \{\delta v\}^T [D_f]\{f(t)\} + \{\delta \beta\}^T [D_q]\{q(t)\} \tag{3.365}$$

so that the generalized force may be put into the form

$$\{\Xi\} = [D_f]\{f(t)\} + [D_q]\{q(t)\} \tag{3.366}$$

and loading matrices for the two-element case given by

$$[D_f] = \begin{bmatrix} \frac{7}{20}\ell_1 & \frac{3}{20}\ell_1 & 0 \\ \frac{1}{20}\ell_1^2 & \frac{1}{30}\ell_1^2 & 0 \\ \frac{3}{20}\ell_1 & \frac{7}{20}(\ell_1+\ell_2) & \frac{3}{20}\ell_2 \\ -\frac{1}{30}\ell_1^2 & \frac{1}{20}(\ell_2^2-\ell_1^2) & \frac{1}{30}\ell_2^2 \\ 0 & \frac{3}{20}\ell_2 & \frac{7}{20}\ell_2 \\ 0 & -\frac{1}{30}\ell_2^2 & -\frac{1}{20}\ell_2^2 \end{bmatrix}$$

$$(3.367)$$

$$[D_q] = \begin{bmatrix} -\frac{1}{2} & -\frac{1}{2} & 0 \\ \frac{1}{12}\ell_1 & -\frac{1}{12}\ell_1 & 0 \\ \frac{1}{2} & 0 & -\frac{1}{2} \\ -\frac{1}{12}\ell_1 & \frac{1}{12}(\ell_1+\ell_2) & -\frac{1}{12}\ell_2 \\ 0 & \frac{1}{2} & \frac{1}{2} \\ 0 & -\frac{1}{12}\ell_2 & \frac{1}{12}\ell_2 \end{bmatrix}$$

The contributions from element 1 are all in the upper-left 4×2 matrix; those from element 2 are in the lower-right 4×2 matrix with overlap in the 2×1 matrix at the center (i.e., the two middle rows of the middle column).

Because the approach is based on the Ritz method, only the geometric boundary conditions need to be satisfied. For a clamped-free beam, this means $v_1 = \beta_1 = 0$, so that the first two rows and columns must be removed from $[M]$ and $[K]$. As for the loading matrices, D_f and D_q, the first two rows must be removed because $\delta v_1 = \delta \beta_1 = 0$. The accuracy of finite elements for beam bending is illustrated in Problem 26.

3.6 Epilogue

In this chapter, we considered the free-vibration analysis and modal representation for flexible structures, along with methods for solving initial-value and forced-response problems associated therewith. Moreover, the approximation techniques of the Ritz method, the Galerkin method, and the finite element method were introduced. This sets the stage for consideration of aeroelastic problems in Chapters 4 and 5. The static-aeroelasticity problem, addressed in Chapter 4, results from interaction of structural and aerodynamic loads. These loads are a subset of those involved in dynamic aeroelasticity, which includes inertial effects. One aspect of dynamic aeroelasticity is flutter, which is discussed in Chapter 5, where it is shown that both the modal representation and the modal approximation methods apply equally well to both types of problems.

Problems

1. By evaluating the appropriate integrals, prove that each function in the following two sets of functions is orthogonal to all other functions in its set over the interval $0 \leq x \leq \ell$:
   (a) $\sin\left(\frac{n\pi x}{\ell}\right)$ for $n = 1, 2, 3, \ldots$
   (b) $\cos\left(\frac{n\pi x}{\ell}\right)$ for $n = 0, 1, 2, \ldots$

   Use of a table of integrals may be helpful.

2. Considering Eq. (3.54), plot the displacement at time $t = 0$ for a varying number of retained modes, showing that as more modes are kept, the shape more closely resembles the initial shape of the string given in Fig. 3.2.

3. Compute the propagation speed of elastic torsional deflections along prismatic, homogeneous, isotropic beams with circular cross sections and made of
   (a) aluminum (2014-T6)
   (b) steel

   Hint: Compare the governing wave equation with that for the uniform-string problem, noting that for beams with a circular cross section, $J = I_p$.
   Answers: (may vary slightly depending on properties used)
   (a) 3,140 m/s
   (b) 3,110 m/s

4. For a uniform string attached between two walls with no external loads, determine the total string deflection $v(x, t)$ for an initial string deflection of zero and an initial transverse velocity distribution given by

$$\frac{\partial v}{\partial t}(x, 0) = V\left[1 - \cos\left(\frac{2\pi x}{\ell}\right)\right]$$

Answer: $v(x, t) = -\dfrac{16V\ell}{\pi^2}\sqrt{\dfrac{m}{T}}\displaystyle\sum_{n=1,3,\ldots}^{\infty}\dfrac{1}{n^2(n^2 - 4)}\sin\left(\dfrac{n\pi x}{\ell}\right)\sin(\omega_n t),$ where

$$\omega_n = \frac{n\pi}{\ell}\sqrt{\frac{T}{m}}$$

5. Consider a uniform string of length ℓ and mass per unit length m that has been stretched between two walls with tension T. Transverse vibration of the string is restrained at its midpoint by a linear spring with spring constant k. The spring is unstretched when the string is undeflected. Write the generalized equation of motion for the ith mode, giving particular attention to the writing of the generalized force Ξ_i. As a check, derive the equation taking into account the spring through the potential energy instead of through the generalized force.

Answer: Letting $\omega_i = \dfrac{i\pi}{\ell}\sqrt{\dfrac{T}{m}}$, we find that the generalized equations of motion are

$$\ddot{\xi}_i + \omega_i^2 \xi_i + \frac{2k}{m\ell}(-1)^{\frac{i-1}{2}} \sum_{j=1,3,\ldots}^{\infty} (-1)^{\frac{j-1}{2}} \xi_j = 0 \qquad i = 1, 3, \ldots, \infty$$

$$\ddot{\xi}_i + \omega_i^2 \xi_i = 0 \qquad\qquad\qquad\qquad\qquad i = 2, 4, \ldots, \infty$$

6. Consider a uniform string of length ℓ with mass per unit length m that has been stretched between two walls with tension T. Until the time $t = 0$, the string is undeflected and at rest. At time $t = 0$, concentrated loads of magnitude $F_0 \sin \Omega t$ are applied at $x = \ell/3$ and $x = 2\ell/3$ in the positive (up) and negative (down) directions, respectively. In addition, a distributed force

$$F = \overline{F}\left[1 - \sin\left(\frac{3\pi x}{\ell}\right)\right]\cos(\Omega t)$$

is applied to the string. What is the total string displacement $v(x, t)$ for time $t > 0$?

Answer: Letting $\omega_n = \dfrac{n\pi}{\ell}\sqrt{\dfrac{T}{m}}$, we find that

$$v(x, t) = \sum_{n=2,4,\ldots}^{\infty} \left\{ C_n \left[\sin(\Omega t) - \frac{\Omega}{\omega_n}\sin(\omega_n t)\right] \sin\left(\frac{n\pi x}{\ell}\right)\right\}$$

$$+ \sum_{n=1,3,\ldots}^{\infty} \left\{ D_n \left[\cos(\Omega t) - \cos(\omega_n t)\right] \sin\left(\frac{n\pi x}{\ell}\right)\right\}$$

where

$$C_n = \frac{2F_0}{m\ell(\omega_n^2 - \Omega^2)}\left[\sin\left(\frac{n\pi}{3}\right) - \sin\left(\frac{2n\pi}{3}\right)\right]$$

$$D_n = \frac{2\overline{F}}{m(\omega_n^2 - \Omega^2)}\left(\frac{2}{n\pi} - \frac{\delta_{n3}}{2}\right)$$

and where the Kronecker symbol $\delta_{ij} = 1$ for $i = j$ and $\delta_{ij} = 0$ for $i \neq j$.

7. Consider a uniform circular rod of length ℓ, torsional rigidity GJ, and mass moment of inertia per unit length ρJ. The beam is clamped at the end $x = 0$, and it has a concentrated inertia I_C at its other end where $x = \ell$.

   (a) Determine the characteristic equation that can be solved for the torsional natural frequencies for the case in which $I_C = \rho J\ell\zeta$, where ζ is a dimensionless parameter.

   (b) Verify that the characteristic equation obtained in part (a) approaches that obtained in the text for the clamped-free uniform rod in torsion as ζ approaches zero.

   (c) Solve the characteristic equation obtained in part (a) for numerical values of the first four eigenvalues, $\alpha_i\ell$, $i = 1, 2, 3$, and 4, when $\zeta = 1$.

(d) Solve the characteristic equation obtained in part (a) for the numerical value of the first eigenvalue, $\alpha_1 \ell$, when $\zeta = 1, 2, 4,$ and 8. Make a plot of the behavior of the lowest natural frequency versus the value of the concentrated inertia. Note that $\alpha_1 \ell$ versus ζ is the same in terms of dimensionless quantities.

Answer:

(a) $\zeta \alpha \ell \tan(\alpha \ell) = 1$

(c, d) Sample result: $\alpha_1 \ell = 0.860334$ for $\zeta = 1$

8. Consider a clamped-free beam undergoing torsion:
   (a) Prove that the free-vibration mode shapes are orthogonal, regardless of whether the beam is uniform.
   (b) Given that the kinetic energy is

$$K = \frac{1}{2} \int_0^\ell \rho I_p \left(\frac{\partial \theta}{\partial t} \right)^2 dx$$

show that K can be written as

$$K = \frac{1}{2} \sum_{i=1}^\infty M_i \dot{\xi}_i^2$$

where M_i is the generalized mass of the ith mode and ξ_i is the generalized coordinate for the ith mode.
   (c) Given that the potential energy is the internal (i.e., strain) energy; that is

$$P = \frac{1}{2} \int_0^\ell \overline{GJ} \left(\frac{\partial \theta}{\partial x} \right)^2 dx$$

show that P can be written as

$$P = \frac{1}{2} \sum_{i=1}^\infty M_i \omega_i^2 \xi_i^2$$

where ω_i is the natural frequency.
   (d) Show that for a uniform beam and for ϕ_i as given in the text, $M_i = \overline{\rho I_p} \ell / 2$ for all i.

9. Consider a uniform free-free beam undergoing torsion:
   (a) Given the mode shapes in the text, find an expression for P in terms of $\overline{GJ}$, ℓ, and the generalized coordinates.
   (b) Given the mode shapes in the text, find an expression for K in terms of $\overline{\rho I_p}$, ℓ, and the time derivatives of the generalized coordinates.
   (c) Substitute results from parts (a) and (b) into Lagrange's equations and identify the resulting generalized masses.

Answer:

(a) $P = \frac{1}{2} \overline{GJ} \sum_{i=1}^\infty \frac{(i\pi)^2}{2\ell} \xi_i^2$

(b) $K = \frac{1}{2} \overline{\rho I_p} \ell \left(\dot{\xi}_0^2 + \frac{1}{2} \sum_{i=1}^\infty \dot{\xi}_i^2 \right)$

(c) $M_0 = \overline{\rho I_p} \ell$; $M_i = \frac{1}{2} \overline{\rho I_p} \ell$ for $i = 1, 2, \ldots$

10. Consider a clamped-free beam undergoing bending:
    (a) Prove that the free-vibration mode shapes are orthogonal, regardless of whether the beam is uniform.
    (b) Given the kinetic energy as

$$K = \frac{1}{2} \int_0^\ell m \left(\frac{\partial v}{\partial t} \right)^2 dx$$

    show that K can be written as

$$K = \frac{1}{2} \sum_{i=1}^{\infty} M_i \dot{\xi}_i^2$$

    where M_i is the generalized mass of the ith mode and ξ_i is the generalized coordinate for the ith mode.
    (c) Given that the potential energy is the internal (i.e., strain) energy; that is

$$P = \frac{1}{2} \int_0^\ell \overline{EI} \left(\frac{\partial^2 v}{\partial x^2} \right)^2 dx$$

    show that P can be written as

$$P = \frac{1}{2} \sum_{i=1}^{\infty} M_i \omega_i^2 \xi_i^2$$

    where ω_i is the natural frequency.
    (d) Show that for a uniform beam and for ϕ_i as given in the text, $M_i = m\ell$ for all i.

11. Consider a uniform beam with the boundary conditions shown in Fig. 3.38 undergoing bending vibration:
    (a) Using the relationships derived in the text, plot the square of characteristic value $(\alpha_1 \ell)^2$, which is proportional to the fundamental frequency, versus κ from 0 to 100. Check your results versus those given in Fig. 3.40.
    (b) Plot the fundamental mode shape for values of κ of $0.01, 0.1, 1, 10$, and 100. Suggestion: use Eq. (3.270). Check your results for $\kappa = 1$ with those given in Fig. 3.39; your results for $\kappa = 100$ will not differ significantly from those in Fig. 3.41, in which $\kappa = 50$.

12. Find the free-vibration frequencies and plot the mode shapes for the first five modes of a beam of length ℓ, having bending stiffness $\overline{EI}$ and mass per unit length m, that is free at its right end, and that has the sliding condition (see Fig. 3.26) at its left end. Normalize the mode shapes to have unit deflection at the free end and determine the generalized mass for the first five modes.
    Answer: $\omega_0 = 0$, $\omega_1 = 5.59332\sqrt{\overline{EI}/(m\ell^4)}$, $\omega_2 = 30.2258\sqrt{\overline{EI}/(m\ell^4)}$, $\omega_3 = 74.6389\sqrt{\overline{EI}/(m\ell^4)}$, $\omega_4 = 138.791\sqrt{\overline{EI}/(m\ell^4)}$; $M_0 = m\ell$ and $M_i = m\ell/4$ for $i = 1, 2, \ldots, \infty$. As a sample of the mode shapes, the first elastic mode is plotted in Fig. 3.48.

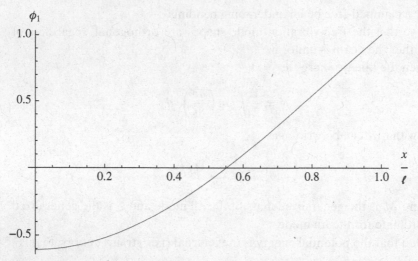

Figure 3.48. First elastic mode shape for sliding-free beam (Note: the "zeroth" mode is a rigid-body translation mode)

13. Consider the beam in Problem 12. Add to it a translational spring restraint at the left end, having spring constant $k = \kappa \overline{EI}/\ell^3$. Find the first three free-vibration frequencies and mode shapes for the cases in which κ takes on values of 0.01, 1, and 100. Plot the mode shapes, normalizing them to have unit deflection at the free end.

    Answer: Sample results: A plot versus κ of $(\alpha_i \ell)^2$ for $i = 1, 2$, and 3 is shown in Fig. 3.49, and the first mode shape for $\kappa = 1$ is shown in Fig. 3.50.

14. Consider a beam that at its left end is clamped and at its right end is pinned with a rigid body attached to it. Let the mass moment of inertia of the rigid body be given by $I_C = \mu m \ell^3$ where C coincides with the pin (i.e., a pivot).

    (a) Find the first two free-vibration frequencies for values of μ equal to 0.01, 0.1, 1, 10, and 100. Comment on the variation of the natural frequencies versus μ.

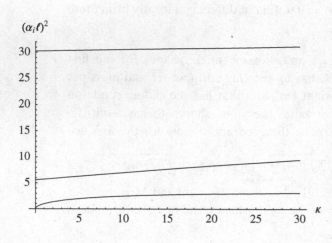

Figure 3.49. Variation versus κ of $(\alpha_i \ell)^2$ for $i = 1, 2$, and 3, for a beam that is free on its right end and has a sliding boundary condition spring-restrained in translation on its left end

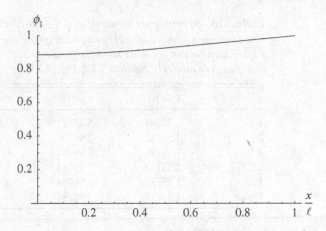

Figure 3.50. First mode shape for a beam that is free on its right end and has a sliding boundary condition spring-restrained in translation on its left end with $\kappa = 1$

(b) Choose any normalization that is convenient and plot the first mode shape for these same values of μ. Comment on the variation of the mode shapes versus μ.

Answer:

(a) Sample result: $\omega_1 = 1.99048\sqrt{\frac{EI}{m\ell^4}}$ for $\mu = 1$.

(b) Sample result: The first mode shape for $\mu = 1$ is shown in Fig. 3.51.

15. Consider a uniform clamped-free beam of length ℓ, bending rigidity $\overline{EI}$, and mass per unit length m. Until time $t = 0$, the beam is undeflected and at rest. At time $t = 0$, a transverse concentrated load of magnitude $F\cos(\Omega t)$ is applied at $x = \ell$.

(a) Write the generalized equations of motion.

(b) Determine the total beam displacement $v(x, t)$ for time $t > 0$.

(c) For the case when $\Omega = 0$, determine the tip displacement of the beam. Ignoring those terms that are time dependent (they would die out in a real beam because of dissipation), plot the tip displacement versus the number of mode shapes retained in the solution up to five modes. Show the static tip deflection from elementary beam theory on the plot. (This part of the problem illustrates how the modal representation can be applied to static-response problems.)

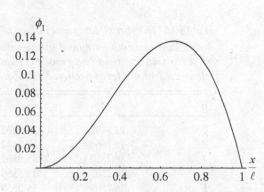

Figure 3.51. First mode shape for a beam that is clamped on its left end and pinned with a rigid body attached on its right end with $\mu = 1$

Table 3.10. *Approximate values of* $\omega_1\sqrt{\frac{m\ell^4}{EI}}$ *for pinned-free beam having a root rotational spring with spring constant of* $\kappa\overline{EI}/\ell$ *using one rigid-body mode (x) and* $n-1$ *clamped-free modes of Section 3.3.4, Eq. (3.258)*

n	$\kappa = 1$	$\kappa = 10$	$\kappa = 100$
1	1.73205	5.47723	17.3205
2	1.55736	2.96790	3.44766
3	1.55730	2.96784	3.44766
4	1.55730	2.96784	3.44766
5	1.55730	2.96784	3.44766
Exact	1.55730	2.96784	3.44766

Answer:
(a) The ith equation is

$$\ddot{\xi}_i + \omega_i^2\xi_i = 2(-1)^{i+1}\frac{F}{m\ell}\cos(\Omega t)$$

(b) With $\phi_i(x)$ given by Eq. (3.258), $\omega_i = (\alpha_i\ell)^2\sqrt{\frac{EI}{m\ell^4}}$, and $\alpha_i\ell$ as given in Table 3.1, we find that

$$v(x,t) = \frac{2F}{m\ell}\sum_{i=1}^{\infty}\frac{(-1)^{i+1}}{\omega_i^2-\Omega^2}\left[\cos(\Omega t)-\cos(\omega_i t)\right]\phi_i(x)$$

(c) The result converges within engineering accuracy to $\frac{F\ell^3}{3EI}$ using only a few terms.

16. Consider a free-free beam with bending stiffness $\overline{EI}$, mass per unit length m, and length ℓ. Applying the Ritz method, write the equations of motion for a system that consists of the beam plus identical rigid bodies attached to the ends, where each body has a moment of inertia I_C and mass m_c. Use as assumed modes those of the exact solution of the free-free beam without the attached bodies, obtained in the text. Note the terms that provide inertial coupling.

17. Consider a pinned-free beam with the rotation about the hinge restrained by a light spring of modulus $\kappa\overline{EI}/\ell$. Use a rigid-body rotation plus the set of clamped-free modes as the assumed modes of the Ritz method. Compare the

Table 3.11. *Approximate values of* $\omega_2\sqrt{\frac{m\ell^4}{EI}}$ *for pinned-free beam having a root rotational spring with spring constant of* $\kappa\overline{EI}/\ell$ *using one rigid-body mode (x) and* $n-1$ *clamped-free modes of Section 3.3.4, Eq. (3.258)*

n	$\kappa = 1$	$\kappa = 10$	$\kappa = 100$
2	22.8402	37.9002	103.173
3	16.2664	19.3632	21.6202
4	16.2512	19.3563	21.6200
5	16.2502	19.3559	21.6200
Exact	16.2501	19.3558	21.6200

Table 3.12. *Approximate values of* $\omega_1\sqrt{\frac{m\ell^4}{EI}}$ *for pinned-free beam having a root rotational spring with spring constant of* $\kappa\overline{EI}/\ell$ *using one rigid-body mode (x) and n − 1 polynomials that satisfy clamped-free beam boundary conditions*

n	$\kappa = 1$	$\kappa = 10$	$\kappa = 100$
1	1.73205	5.47723	17.3205
2	1.55802	2.97497	3.46064
3	1.55730	2.96784	3.44768
4	1.55730	2.96784	3.44766
Exact	1.55730	2.96784	3.44766

results for the first two modes using a varying number of terms for $\kappa = 1, 10,$ and 100.
Answer: See Tables 3.10 and 3.11.

18. Repeat Problem 17 using a set of polynomial admissible functions. Use one rigid-body mode (x) and a varying number of polynomials that satisfy all the boundary conditions of a clamped-free beam.
    Answer: See Tables 3.12 and 3.13.

19. Consider a clamped-free beam of length ℓ for which the mass per unit length and bending stiffness vary according to

$$m = m_0 \left(1 - \frac{x}{\ell} + \mu\frac{x}{\ell}\right)$$

$$\overline{EI} = \overline{EI}_0 \left(1 - \frac{x}{\ell} + \kappa\frac{x}{\ell}\right)$$

Using the comparison functions in Eq. (3.325), apply the Ritz method to determine approximate values for the first three natural frequencies, varying the number of terms from one to five. Let $\mu = \kappa = 1/2$.
Answer: See Table 3.14.

20. Rework Problem 19 using the Ritz method and the set of polynomial admissible functions $(x/\ell)^{i+1}$, $i = 1, 2, \ldots, n$
    Answer: See Table 3.15.

Table 3.13. *Approximate values of* $\omega_2\sqrt{\frac{m\ell^4}{EI}}$ *for pinned-free beam having a root rotational spring with spring constant of* $\kappa\overline{EI}/\ell$ *using one rigid-body mode (x) and n − 1 polynomials that satisfy clamped-free beam boundary conditions*

n	$\kappa = 1$	$\kappa = 10$	$\kappa = 100$
2	24.8200	41.1049	111.743
3	16.4047	19.7070	22.2338
4	16.2508	19.3565	21.6208
Exact	16.2501	19.3558	21.6200

Table 3.14. *Approximate values of* $\omega_i \sqrt{\frac{m_0 \ell^4}{EI_0}}$ *for a tapered, clamped-free beam based on the Ritz method with n polynomials that satisfy all the boundary conditions of a clamped-free beam*

n	$\omega_1 \sqrt{\frac{m_0 \ell^4}{EI_0}}$	$\omega_2 \sqrt{\frac{m_0 \ell^4}{EI_0}}$	$\omega_3 \sqrt{\frac{m_0 \ell^4}{EI_0}}$
1	4.36731	–	–
2	4.31571	24.7653	–
3	4.31517	23.5267	69.8711
4	4.31517	23.5199	63.2441
5	4.31517	23.5193	63.2415
Exact	4.31517	23.5193	63.1992

21. Rework Problem 19 using Eq. (3.329) with $f = 0$ as the equation of motion and the set of polynomial comparison functions $(x/\ell)^{i+1}$, $i = 1, 2, \ldots, n$.
    Answer: See Table 3.16.

22. Consider a clamped-free beam to which is attached at spanwise location $x = \ell r$ a particle of mass $\mu m \ell$. Using a two-term Ritz approximation based on the functions in Eq. (3.325), plot the approximate value for the fundamental natural frequency as a function of r for $\mu = 1$.
    ans.: See Fig. 3.52.

23. Consider a clamped-free beam undergoing coupled bending and torsion. Set up an approximate solution based on the Ritz method for the dimensionless frequency parameter

$$\lambda^2 = \frac{m \ell^4 \omega^2}{EI}$$

using the uncoupled bending and torsional mode shapes as assumed modes, and with the parameters $\overline{\rho I_p} = 0.01 m \ell^2$, $md^2 = 0.25 \overline{\rho I_p}$, $K^2 = 0.25 \overline{GJ}\ \overline{EI}$, and $\overline{GJ/EI} = 5$. Answer these questions: How do the signs of d and K affect the frequencies? How do they affect the predicted mode shapes?

Table 3.15. *Approximate values of* $\omega_i \sqrt{\frac{m_0 \ell^4}{EI_0}}$ *for a tapered, clamped-free beam based on the Ritz method with n terms of the form* $(x/\ell)^{i+1}$, $i = 1, 2, \ldots, n$

n	$\omega_1 \sqrt{\frac{m_0 \ell^4}{EI_0}}$	$\omega_2 \sqrt{\frac{m_0 \ell^4}{EI_0}}$	$\omega_3 \sqrt{\frac{m_0 \ell^4}{EI_0}}$
1	5.07093	–	–
2	4.31883	33.8182	–
3	4.31732	23.6645	110.529
4	4.31523	23.6640	64.8395
5	4.31517	23.5226	64.7821
Exact	4.31517	23.5193	63.1992

Table 3.16. *Approximate values of $\omega_i \sqrt{\frac{m_0 \ell^4}{EI_0}}$ for a tapered, clamped-free beam based on the Galerkin method applied to Eq. (3.329) with n terms of the form $(x/\ell)^{i+1}$, $i = 1, 2, \ldots, n$*

n	$\omega_1 \sqrt{\frac{m_0 \ell^4}{EI_0}}$	$\omega_2 \sqrt{\frac{m_0 \ell^4}{EI_0}}$	$\omega_3 \sqrt{\frac{m_0 \ell^4}{EI_0}}$
1	7.88811	–	–
2	4.45385	54.5221	–
3	4.19410	24.3254	175.623
4	4.33744	21.4784	67.1265
5	4.31379	23.8535	53.6214
Exact	4.31517	23.5193	63.1992

24. Repeat Problem 23 using an appropriate power series for bending and for torsion.

25. Develop a finite-element solution for the static twist of a clamped-free beam in torsion, accounting for linearly varying $\overline{GJ}(x)$ within each element. Compare results for the tip rotation caused by twisting, with identical loading and properties (i.e., $\overline{GJ}(0) = \overline{GJ}_0 = 2\overline{GJ}(\ell)$, $r(x, t) = r = $ const.). Note that the results in Section 3.5.3 approximate the linearly varying $\overline{GJ}$ as piecewise constant within elements, whereas here you are to assume piecewise linearly varying $\overline{GJ}$ within elements.

Answer: The results do not change; see Table 3.9.

26. Set up a finite-element solution for the dimensionless natural frequencies of a beam in bending $m(0)\ell^4\omega^2/[\overline{EI}(0)]$ from Section 3.5.3, accounting for linearly

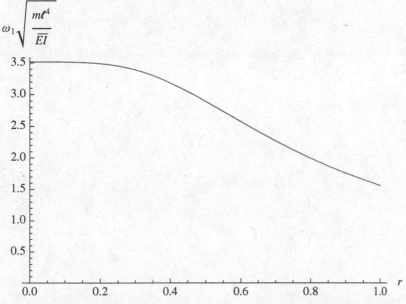

Figure 3.52. Approximate fundamental frequency for a clamped-free beam with a particle of mass $m\ell$ attached at $x = r\ell$

Table 3.17. *Finite-element results for the natural frequencies of a beam in bending with linearly varying $\overline{EI}(x)$, such that $\overline{EI}(0) = \overline{EI}_0 = 2\overline{EI}(\ell)$ and values of $\overline{EI}$ are taken as linear within each element*

n	$\omega_1\sqrt{\frac{m_0\ell^4}{EI_0}}$	$\omega_2\sqrt{\frac{m_0\ell^4}{EI_0}}$	$\omega_3\sqrt{\frac{m_0\ell^4}{EI_0}}$
1	4.31883	33.8182	–
2	4.31654	23.6457	75.9255
3	4.31549	23.5835	63.8756
4	4.31528	23.5430	63.6528
5	4.31522	23.5296	63.4128
6	4.31519	23.5244	63.3088
Exact	4.31517	23.5193	63.1992

varying $\overline{EI}(x)$ and $m(x)$ within each element. Compare $[M]$ and $[K]$ matrices obtained for the case developed in the text (i.e., piecewise constant $\overline{EI}$ and m within elements) with those obtained for linearly varying within elements. Tabulate dimensionless frequencies for the first three modes, assuming elements of constant length, $\overline{EI}(\ell) = 0.5\overline{EI}(0)$, and $m(\ell) = 0.5m(0)$.

Answer: See Table 3.17.

4 Static Aeroelasticity

I discovered that with increasing load, the angle of incidence at the wing tips increased perceptibly. It suddenly dawned on me that this increasing angle of incidence was the cause of the wing's collapse, as logically the load resulting from the air pressure in a steep dive would increase faster at the wing tips than at the middle. The resulting torsion caused the wings to collapse under the strain of combat maneuvers.
—A. H. G. Fokker in *The Flying Dutchman*, Henry Holt and Company, 1931

The field of static aeroelasticity is the study of flight-vehicle phenomena associated with the interaction of aerodynamic loading induced by steady flow and the resulting elastic deformation of the lifting-surface structure. These phenomena are characterized as being insensitive to the rates and accelerations of the structural deflections. There are two classes of design problems that are encountered in this area. The first and most common to all flight vehicles is the effects of elastic deformation on the airloads, as well as effects of airloads on the elastic deformation, associated with normal operating conditions. These effects can have a profound influence on performance, handling qualities, flight stability, structural-load distribution, and control effectiveness. The second class of problems involves the potential for static instability of the lifting-surface structure to result in a catastrophic failure. This instability is often termed "divergence" and it can impose a limit on the flight envelope. Simply stated, divergence occurs when a lifting surface deforms under aerodynamic loads in such a way as to increase the applied load, and the increased load deflects the structure further—eventually to the point of failure. Such a failure is not simply the result of a load that is too large for the structure as designed; instead, the aerodynamic forces actually interact with the structure to create a loss of effective stiffness. This phenomenon is explored in more detail in this chapter.

The material presented in this chapter is an introduction to some of these static aeroelastic phenomena. To illustrate clearly the mechanics of these problems and yet maintain a low level of mathematical complexity, relatively simple configurations are considered. The first items treated are rigid aerodynamic models that are elastically mounted in a wind-tunnel test section; such elastic mounting is characteristic of most load-measurement systems. The second aeroelastic configuration to be treated is

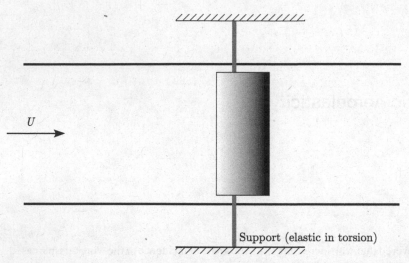

Figure 4.1. Planform view of a wind-tunnel model on a torsionally elastic support

a uniform elastic lifting surface of finite span. Its static aeroelastic properties are similar to those of most lifting surfaces on conventional flight vehicles.

4.1 Wind-Tunnel Models

In this section, we consider three types of mounting for wind-tunnel models: wall-mounted, sting-mounted, and strut-mounted. Expressions for the aeroelastic pitch deflections are developed for these simple models that, in turn, lead to a cursory understanding of the divergence instability. Finally, we briefly return to the wall-mounted model in this section to consider the qualitatively different phenomenon of aileron reversal. All of these wing models are assumed to be rigid and two-dimensional. That is, the airfoil geometry is independent of spanwise location, and the span is sufficiently large that the lift and pitching moment do not depend on a spanwise coordinate.

4.1.1 Wall-Mounted Model

Consider a rigid, spanwise-uniform model of a wing that is mounted to the side walls of a wind tunnel in such a way as to allow the wing to pitch about the support axis, as illustrated in Fig. 4.1. The support is flexible in torsion, which means that it restricts the pitch rotation of the wing in the same way as a rotational spring would. We denote the rotational stiffness of the support by k, as shown in Fig. 4.2. If we assume the body to be pivoted about its support O, located at a distance x_O from the leading edge, moment equilibrium requires that the sum of all moments about O must equal zero. Thus

$$M_{ac} + L(x_O - x_{ac}) - W(x_O - x_{cg}) - k\theta = 0 \qquad (4.1)$$

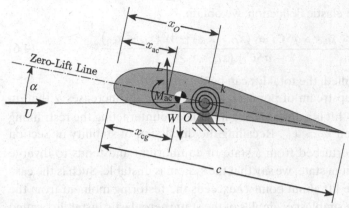

Figure 4.2. Airfoil for wind-tunnel model

Were the support rigid, the angle of attack would be α_r, positive nose-up. The elastic part of the pitch angle is denoted by θ, which is also positive nose-up. The wing angle of attack is then $\alpha \equiv \alpha_r + \theta$. In anticipation of using linear aerodynamics, we assume the angle of attack, α, to be a small angle, such that $\sin(\alpha) \approx \alpha$ and $\cos(\alpha) \approx 1$. It also is necessary to restrict the analysis to "thin" airfoils (i.e., small thickness to chord and small camber). The treatment herein is restricted to incompressible flow, but compressibility effects may be taken into account by means of Prandtl-Glauert corrections to the airfoil coefficients. For this, the freestream Mach number must remain less than roughly 0.8 to avoid transonic effects.

For linear aerodynamics, the lift for a rigid support is simply

$$L_{\text{rigid}} = qSC_{L_\alpha}\alpha_r \tag{4.2}$$

whereas the lift for an elastic support is

$$L = qSC_{L_\alpha}(\alpha_r + \theta) \tag{4.3}$$

where $q = \frac{1}{2}\rho_\infty U^2$ is the freestream dynamic pressure (i.e., in the far field—often denoted by q_∞), U is the freestream air speed, ρ_∞ is the freestream air density, S is the planform area, and C_{L_α} is the wing lift-curve slope. Note that $L \neq L_{\text{rigid}}$; for positive θ, $L > L_{\text{rigid}}$. We can express the moment of aerodynamic forces about the aerodynamic center as

$$M_{\text{ac}} = qScC_{M\text{ac}} \tag{4.4}$$

If the angle of attack is small, $C_{M\text{ac}}$ can be regarded as a constant. Note here that linear aerodynamics implies that the lift-curve slope C_{L_α} is a constant. A further simplification may be that $C_{L_\alpha} = 2\pi$ in accordance with two-dimensional thin-airfoil theory. If experimental data or results from computational fluid dynamics provide an alternative value, then it should be used.

Using Eqs. (4.3) and (4.4), the equilibrium equation, Eq. (4.1), can be expanded as

$$qScC_{M\text{ac}} + qSC_{L_\alpha}(\alpha_r + \theta)(x_O - x_{\text{ac}}) - W(x_O - x_{\text{cg}}) = k\theta \tag{4.5}$$

Solving Eq. (4.5) for the elastic deflection, we obtain

$$\theta = \frac{qScC_{Mac} + qSC_{L_\alpha}\alpha_r(x_O - x_{ac}) - W(x_O - x_{cg})}{k - qSC_{L_\alpha}(x_O - x_{ac})} \qquad (4.6)$$

When α_r and q are specified, the total lift can be determined.

When the lift acts upstream of point O, an increase in lift increases α that, in turn, increases lift. Thus, lift is a destabilizing influence counteracting the restraining action of the spring when $x_O > x_{ac}$. Recalling the discussion of stability in Section 2.5, when a system is perturbed from a state of equilibrium and tends to diverge further from its equilibrium state, we say that the system is unstable. Such is the case when the moment of the lift about point O exceeds the restoring moment from the spring. This is one of the simplest examples of the static aeroelastic instability called "divergence." Now, from Eq. (4.6), we see that the aerodynamic center is forward of the support point O when $x_{ac} < x_O$, making it possible for the denominator to vanish or for θ to blow up when q is sufficiently large. The denominator of the expression for θ is a sort of effective stiffness, which decreases as q increases. When the denominator vanishes, divergence occurs. The divergence dynamic pressure—or dynamic pressure at which divergence occurs—is then denoted by

$$q_D = \frac{k}{SC_{L_\alpha}(x_O - x_{ac})} \qquad (4.7)$$

From this, the divergence speed—or the air speed at which divergence occurs—can be found as

$$U_D = \sqrt{\frac{2k}{\rho_\infty SC_{L_\alpha}(x_O - x_{ac})}} \qquad (4.8)$$

It is evident that when the aerodynamic center is coincident with the pivot, so that $x_O = x_{ac}$, the divergence dynamic pressure becomes infinite. Also, when the aerodynamic center is aft of the pivot so that $x_O < x_{ac}$, the divergence dynamic pressure becomes negative. Because for physical reasons dynamic pressure must be positive and finite, it is clear in either case that divergence is impossible.

To further pursue the character of this instability, consider the case of a symmetric airfoil ($C_{Mac} = 0$). Furthermore, let $x_O = x_{cg}$ so that the weight term drops out of the equation for θ. From Eq. (4.7), we can let $k = q_D SC_{L_\alpha}(x_O - x_{ac})$; therefore, θ can be written simply as

$$\theta = \frac{\alpha_r}{\frac{q_D}{q} - 1} \qquad (4.9)$$

The lift is proportional to $\alpha_r + \theta$. Thus, the change in lift divided by the rigid lift is given by

$$\frac{\Delta L}{L_{rigid}} = \frac{\theta}{\alpha_r} = \frac{\frac{q}{q_D}}{1 - \frac{q}{q_D}} \qquad (4.10)$$

Both θ and $\Delta L/L_{rigid}$ clearly approach infinity as $q \to q_D$. Indeed, a plot of the latter is given in Fig. 4.3 and shows the large change in lift caused by the aeroelastic effect. The lift evidently starts from its "rigid" value—that is, the value it would have were

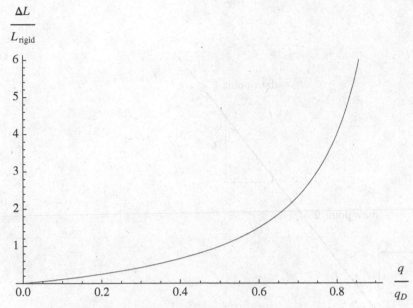

Figure 4.3. Relative change in lift due to aeroelastic effect

the support rigid—and increases to infinity as $q \to q_D$. However, remember that there are limitations on the validity of both expressions. Namely, the lift will not continue to increase as stall is encountered. Moreover, because the structure will not tolerate infinite deformation, failure takes place at some finite value of θ—generally at a dynamic pressure well below the divergence dynamic pressure.

When the system parameters are within the bounds of validity for linear theory, another fascinating feature of this problem emerges. We can invert the expression for θ to obtain

$$\frac{1}{\theta} = \frac{q_D}{\alpha_r} \left(\frac{1}{q} - \frac{1}{q_D} \right) \qquad (4.11)$$

making it evident that $1/\theta$ is proportional to $1/q$ (Fig. 4.4). Therefore, for a model of this type, only two data points are needed to extrapolate the line down and to the left until it intercepts the $1/q$ axis at a distance $1/q_D$ from the origin. As shown in the figure, the slope of this line also can be used to estimate q_D. The form of this plot is of great practical value because estimates of q_D can be extrapolated from data taken at speeds far below the divergence speed. This means that q_D can be estimated even when the values of the model parameters are not precisely known, thereby circumventing the need to risk destruction of the model by testing all the way up to the divergence boundary.

4.1.2 Sting-Mounted Model

A second configuration of potential interest is a rigid model mounted on an elastic sting. A simplified version of this kind of model is shown in Figs. 4.5 through 4.7, in which the sting is modeled as a uniform, elastic, clamped-free beam with bending

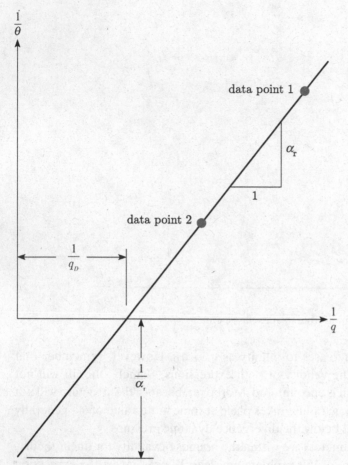

Figure 4.4. Plot of $1/\theta$ versus $1/q$

stiffness $\overline{EI}$ and length λc, where λ is a dimensionless parameter. The model is mounted in such a way as to have angle of attack of α_r when the beam is undeformed. Thus, as before, $\alpha = \alpha_r + \theta$, where θ is the nose-up rotation of the wing resulting from bending of the sting, as shown in Fig. 4.6. Also in Fig. 4.6, we denote the tip deflection of the beam as δ, although we do not need it for this analysis. Note the equal and opposite directions on the force F_0 and moment M_0 at the trailing edge of the wing in Fig. 4.7 versus at the tip of the sting in Fig. 4.6.

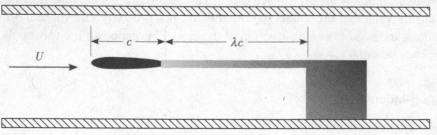

Figure 4.5. Schematic of a sting-mounted wind-tunnel model

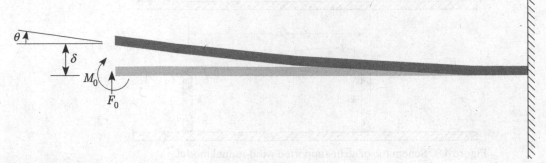

Figure 4.6. Detailed view of the clamped-free beam

From superposition, we can deduce the total bending slope at the tip of the sting as the sum of contributions from the tip force F_0 and tip moment M_0, denoted by θ_F and θ_M, respectively, so that

$$\theta = \theta_F + \theta_M \tag{4.12}$$

From elementary beam theory, these constituent parts can be written as

$$\theta_F = \frac{F_0(\lambda c)^2}{2\overline{EI}}$$
$$\theta_M = \frac{M_0(\lambda c)}{\overline{EI}} \tag{4.13}$$

so that

$$F_0 = \frac{2\overline{EI}\,\theta_F}{(\lambda c)^2}$$
$$M_0 = \frac{\overline{EI}\,\theta_M}{\lambda c} \tag{4.14}$$

Two static aeroelastic equilibrium equations now can be written for the determination of θ_F and θ_M. Using Eqs. (4.3) and (4.4) for the lift and pitching moment,

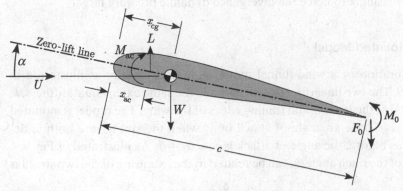

Figure 4.7. Detailed view of the sting-mounted wing

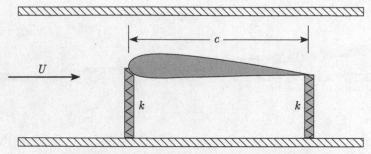

Figure 4.8. Schematic of strut-supported wind-tunnel model

the force equilibrium equation can be written as

$$qSC_{L_\alpha}(\alpha_r + \theta_F + \theta_M) - W - F_0 = 0 \tag{4.15}$$

and the sum of moments about the trailing edge yields

$$qScC_{Mac} + qSC_{L_\alpha}(\alpha_r + \theta_F + \theta_M)(c - x_{ac}) - W(c - x_{cg}) - M_0 = 0 \tag{4.16}$$

Substitution of Eqs. (4.14) into Eqs. (4.15) and (4.16), simultaneous solution for θ_F and θ_M, and use of Eq. (4.12) yields

$$\theta = \frac{\dfrac{W(\lambda + 2 - 2r_{cg}) - 2qSC_{Mac}}{qSC_{L_\alpha}} - \alpha_r(\lambda + 2 - 2r_{ac})}{\lambda + 2 - 2r_{ac} - \dfrac{2\overline{EI}}{\lambda c^2 qSC_{L_\alpha}}} \tag{4.17}$$

where $r_{ac} = x_{ac}/c$ and $r_{cg} = x_{cg}/c$. Here again, the condition for divergence can be obtained by setting the denominator to zero, so that

$$q_D = \frac{2\overline{EI}}{c^2 S\lambda(\lambda + 2 - 2r_{ac})C_{L_\alpha}} \tag{4.18}$$

However, unlike the previous example, we cannot make the divergence dynamic pressure infinite or negative (thereby making divergence mathematically impossible) by choice of configuration parameters because $x_{ac}/c \leq 1$. For a given wing configuration, we are left only with the possibility of increasing the sting's bending stiffness or decreasing λ to make the divergence dynamic pressure larger.

4.1.3 Strut-Mounted Model

A third configuration of a wind-tunnel mount is a strut system, as illustrated in Figs. 4.8 and 4.9. The two linearly elastic struts have the same extensional stiffness, k, and are mounted at the leading and trailing edges of the wing. The model is mounted in such a way as to have an angle of attack of α_r when the springs are both undeformed. Thus, as before, the angle of attack is $\alpha = \alpha_r + \theta$. As illustrated in Fig. 4.9, the elastic part of the pitch angle, θ, can be related to the extension of the two struts as

$$\theta = \frac{\delta_1 - \delta_2}{c} \tag{4.19}$$

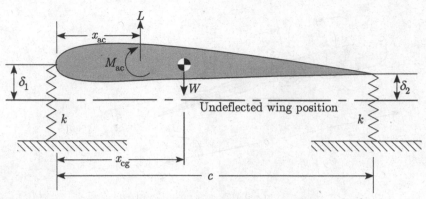

Figure 4.9. Cross section of strut-supported wind-tunnel model

The sum of the forces in the vertical direction shows that

$$L - W - k(\delta_1 + \delta_2) = 0 \qquad (4.20)$$

The sum of the moments about the trailing edge yields

$$M_{ac} + L(c - x_{ac}) - W(c - x_{cg}) - kc\delta_1 = 0 \qquad (4.21)$$

Again, using Eqs. (4.3) and (4.4) for the lift and pitching moment, the simultaneous solution of the force and moment equations yields

$$\theta = \frac{\alpha_r \left(1 - 2\frac{x_{ac}}{c}\right) + 2\frac{C_{Mac}}{C_{L_\alpha}} - \frac{W}{qSC_{L_\alpha}}\left(1 - 2\frac{x_{cg}}{c}\right)}{\frac{kc}{qSC_{L_\alpha}} - \left(1 - 2\frac{x_{ac}}{c}\right)} \qquad (4.22)$$

As usual, the divergence condition is indicated by the vanishing of the denominator, so that

$$q_D = \frac{kc}{SC_{L_\alpha}\left(1 - 2\frac{x_{ac}}{c}\right)} \qquad (4.23)$$

It is evident for this problem as specified that when the aerodynamic center is in front of the mid-chord (as it is in subsonic flow), the divergence condition cannot be eliminated. However, divergence can be eliminated if the leading-edge spring stiffness is increased relative to that of the trailing-edge spring. This is left as an exercise for readers (see Problem 5).

4.1.4 Wall-Mounted Model for Application to Aileron Reversal

Before leaving the wind-tunnel-type models discussed so far in this chapter, we consider the problem of aileron reversal. "Aileron reversal" is the reversal of the aileron's expected response due to structural deformation of the wing. For example, wing torsional flexibility can cause ailerons to gradually lose their effectiveness as dynamic pressure increases; beyond a certain dynamic pressure that we call the "reversal dynamic pressure," they start to function in a manner that is opposite to their intended purpose. The primary danger posed by the loss of control effectiveness

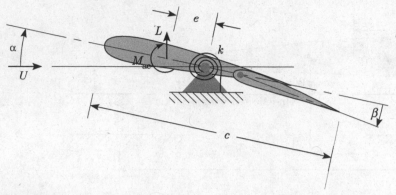

Figure 4.10. Schematic of the airfoil section of a flapped two-dimensional wing in a wind tunnel

is that the pilot cannot control the aircraft in the usual way. There are additional concerns for aircraft, the missions of which depend on their being highly maneuverable. For example, when control effectiveness is lost, the pilot may not be able to count on the aircraft's ability to execute evasive maneuvers. This loss in control effectiveness and eventual reversal is the focus of this section.

Consider the airfoil section of a flapped two-dimensional wing, shown in Fig. 4.10. Similar to the model discussed in Section 4.1.1, the wing is pivoted and restrained by a rotational spring with spring constant k. The main differences are that (1) a trailing-edge flap is added such that the flap angle β can be arbitrarily set by the flight-control system; and (2) we need not consider gravity to illustrate this phenomenon, so the weight is not shown in the figure. Moment equilibrium for this system about the pivot requires that

$$M_{\mathrm{ac}} + eL = k\theta \tag{4.24}$$

The lift and pitching moment for a two-dimensional wing can be written as before; namely

$$L = qSC_L$$
$$M_{\mathrm{ac}} = qcSC_{Mac} \tag{4.25}$$

When $\beta \neq 0$, the effective camber of the airfoil changes, inducing changes in both lift and pitching moment. For a linear theory, both α and β should be small angles, so that

$$C_L = C_{L_\alpha}\alpha + C_{L_\beta}\beta$$
$$C_{Mac} = C_{M_0} + C_{M_\beta}\beta \tag{4.26}$$

where, as before, the angle of attack is $\alpha = \alpha_{\mathrm{r}} + \theta$. Note that $C_{M_\beta} < 0$; for convenience, we assume a symmetric airfoil ($C_{M_0} = 0$).

Note that we may most directly determine the divergence dynamic pressure by writing the equilibrium equation without the inhomogeneous terms; that is

$$(k - eqSC_{L_\alpha})\theta = 0 \tag{4.27}$$

A nontrivial solution exists when the coefficient of θ vanishes, yielding the divergence dynamic pressure as

$$q_D = \frac{k}{eSC_{L_\alpha}} \tag{4.28}$$

Clearly, the divergence dynamic pressure is unaffected by the aileron.

Conversely, the response is significantly affected by the aileron, as we now show. We can solve the response problem by substituting Eqs. (4.25) into the moment-equilibrium equation, Eq. (4.24), making use of Eqs. (4.26), and determining θ to be

$$\theta = \frac{qS\left[eC_{L_\alpha}\alpha_r + \left(eC_{L_\beta} + cC_{M_\beta}\right)\beta\right]}{k - eqSC_{L_\alpha}} \tag{4.29}$$

We see that because of the flexibility of the model in pitch (representative of torsional flexibility in a wing), θ is a function of β. We then find the lift as follows:

1. Substitute Eq. (4.29) into $\alpha = \alpha_r + \theta$ to obtain α.
2. Substitute α into the first of Eqs. (4.26) to obtain the lift coefficient.
3. Finally, substitute the lift coefficient into the first of Eqs. (4.25) to obtain an expression for the aeroelastic lift:

$$L = \frac{qS\left[C_{L_\alpha}\alpha_r + C_{L_\beta}\left(1 + \frac{cqSC_{L_\alpha}C_{M_\beta}}{kC_{L_\beta}}\right)\beta\right]}{1 - \frac{eqSC_{L_\alpha}}{k}} \tag{4.30}$$

It is evident from the two terms in the coefficient of β in this expression that lift is a function of β in two counteracting ways. Ignoring the effect of the denominator, we see that the first term in the numerator that multiplies β is purely aerodynamic and leads to an *increase* in lift with β because of a change in the effective camber. The second term is aeroelastic. Recalling that $C_{M_\beta} < 0$, we see that as β is increased, the effective change in the camber also induces a nose-down pitching moment that—because the model is flexible in pitch—tends to decrease θ and in turn *decrease* lift. At low speed, the purely aerodynamic increase in lift overpowers the aeroelastic tendency to decrease the lift, so that the lift indeed increases with β (and the aileron works as advertised). However, as dynamic pressure increases, the aeroelastic effect becomes stronger; there is a point at which the net rate of change of lift with respect to β vanishes so that

$$\frac{\partial L}{\partial \beta} = 0 = \frac{qSC_{L_\beta}\left(1 + \frac{cqSC_{L_\alpha}C_{M_\beta}}{kC_{L_\beta}}\right)}{1 - \frac{eqSC_{L_\alpha}}{k}} \tag{4.31}$$

Thus, we find that the dynamic pressure at which the reversal occurs is

$$q_R = -\frac{kC_{L_\beta}}{cSC_{L_\alpha}C_{M_\beta}} \tag{4.32}$$

Notice that because $C_{M_\beta} < 0$, $q_R > 0$. Obviously, a stiffer k gives a higher reversal speed, and a model that is rigid in pitch (analogous to a torsionally rigid wing) will not undergo reversal. For dynamic pressures above q_R (but still below the divergence dynamic pressure), a positive β will actually decrease the lift.

Now let us consider the effect of both numerator and denominator. As discussed previously, the divergence dynamic pressure also can be found by setting the denominator of L or θ equal to zero, resulting in the same expression for q_D as found in Eq. (4.27). Equations (4.27) and (4.32) can be used to simplify the expression for the lift in Eq. (4.30) to obtain

$$L = \frac{qS\left[C_{L_\alpha}\alpha_{\mathrm{r}} + C_{L_\beta}\left(1 - \frac{q}{q_R}\right)\beta\right]}{1 - \frac{q}{q_D}} \tag{4.33}$$

It is clear from this expression that the coefficient of β can be positive, negative, or zero. Thus, a positive β could increase the lift, decrease the lift, or not change the lift at all. The aileron's lift efficiency, η, can be thought of as the aeroelastic (i.e., actual) change in lift per unit change in β divided by the change in lift per unit change in β that would result were the model not flexible in pitch; that is

$$\eta = \frac{\text{change in lift per unit change in } \beta \text{ for elastic wing}}{\text{change in lift per unit change in } \beta \text{ for rigid wing}}$$

Using this, we can easily find that

$$\eta = \frac{1 - \frac{q}{q_R}}{1 - \frac{q}{q_D}} \tag{4.34}$$

which implies that the wing will remain divergence-free and control efficiency will not be lost as long as $q < q_D \leq q_R$. Obviously, were the model rigid in pitch, both q_D and q_R would become infinite and $\eta = 1$.

Thinking unconventionally for the moment, let us allow the possibility of $q_R \ll q_D$. This will result in aileron reversal at a low speed, of course. Although the aileron now works opposite to the usual way at most operational speeds of the aircraft, this type of design should not be ruled out on these grounds alone. Active flight-control systems certainly can compensate for this. Moreover, we can obtain considerably more (negative) lift for positive β in this unusual regime than positive lift for positive β in the more conventional setting. This concept is a part of the design of the Kaman "servo-flap rotor," the blades of which have trailing-edge flaps that flap up for increased lift. It also may have important implications for the design of highly maneuverable aircraft. Exactly what other potential advantages and disadvantages exist from following this strategy—particularly in this era of composite materials, smart structures, and active controls—is not presently known and is the subject of current research.

We revisit this problem in Section 4.2.5 from the point of view of a flexible beam model for the wing.

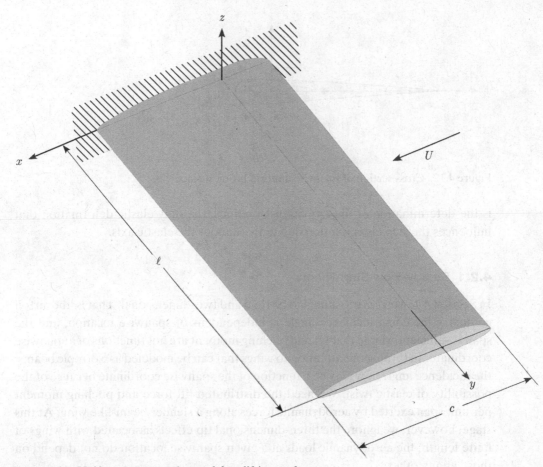

Figure 4.11. Uniform unswept clamped-free lifting surface

4.2 Uniform Lifting Surface

So far, our aeroelastic analyses focused on rigid wings with a flexible support. These idealized configurations provide insight into the aeroelastic stability and response, but practical analyses must take into account flexibility of the lifting surface. That being the case, in this section, we address flexible wings, albeit with simplified structural representation.

Consider an unswept uniform elastic lifting surface as illustrated in Figs. 4.11 and 4.12. The lifting surface is modeled as a beam and, in keeping with historical practice in the field of aeroelasticity, the spanwise coordinate along the elastic axis is denoted by y. The beam is presumed to be built in at the root (i.e., $y = 0$, to represent attachment to a wind-tunnel wall or a fuselage) and free at the tip (i.e., $y = \ell$). The y axis corresponds to the elastic axis, which may be defined as the line of effective shear centers, assumed here to be straight. Recall that for isotropic beams, a transverse force applied at any point along this axis results in bending with no elastic torsional rotation about the axis. This axis is also the axis of twist in response to a pure twisting moment applied to the wing. Because the primary concern here

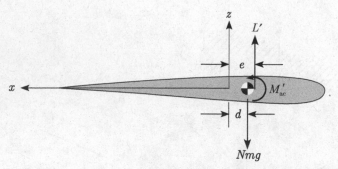

Figure 4.12. Cross section of spanwise uniform lifting surface

is the determination of the airload distributions, the only elastic deformation that influences these loads is rotation due to twist about the elastic axis.

4.2.1 Steady-Flow Strip Theory

In Section 4.1, wings are assumed to be rigid and two-dimensional. That is, the airfoil geometry including incidence angle is independent of spanwise location, and the span is sufficiently large that lift and pitching moment are not functions of a spanwise coordinate. In turning our attention to wings that can be modeled as isotropic beams, the incidence angle now may be a function of the spanwise coordinate because of the possibility of elastic twist. We need the distributed lift force and pitching moment per unit span exerted by aerodynamic forces along a slender beam-like wing. At this stage, however, we ignore the three-dimensional tip effects associated with wings of finite length; the aerodynamic loads at a given spanwise location do not depend on those at any other.

The total applied, distributed, twisting moment per unit span about the elastic axis is denoted as $M'(y)$, which is positive leading-edge-up and given by

$$M' = M'_{ac} + eL' - Nmgd \qquad (4.35)$$

where L' and M'_{ac} are the distributed spanwise lift and pitching moment (i.e., the lift and pitching moment per unit length), mg is the spanwise weight distribution (i.e., the weight per unit length), and N is the "normal load factor" for the case in which the wing is level (i.e., the z axis is directed vertically upward). Thus, N can be written as

$$N = \frac{L}{W} = 1 + \frac{A_z}{g} \qquad (4.36)$$

where A_z is the z component of the wing's inertial acceleration, W is the total weight of the aircraft, and L is the total lift.

The distributed aerodynamic loads can be written in coefficient form as

$$L' = qcc_\ell$$
$$M'_{ac} = qc^2 c_{mac} \qquad (4.37)$$

where the freestream dynamic pressure, q, is

$$q = \frac{1}{2}\rho_\infty U^2 \qquad (4.38)$$

Note that the sectional lift c_ℓ and moment c_{mac} coefficients are written here in lower case to distinguish them from lift and pitching moment coefficients for a two-dimensional wing, which are normally written in upper case. Finally, the primes are included with L', M', and M'_{ac} to reflect that these are distributed quantities (i.e., per unit span).

The sectional lift and pitching-moment coefficients can be related to the angle of attack α by an appropriate aerodynamic theory as some functions $c_\ell(\alpha)$ and $c_{mac}(\alpha)$, where the functional relationship generally involves integration over the planform. To simplify the calculation, the wing can be broken up into spanwise segments of infinitesimal length, where the local lift and pitching moment can be estimated from two-dimensional theory. This theory, commonly known as "strip theory," frequently uses a table for efficient calculation. Here, however, for small values of α, we may use an even simpler form in which the lift–curve slope is assumed to be a constant along the span, so that

$$c_\ell(y) = a\alpha(y) \qquad (4.39)$$

where a denotes the constant sectional lift–curve slope, and the sectional-moment coefficient $c_{mac}(\alpha)$ is assumed to be a constant along the span.

The angle of attack is represented by two components. The first is a rigid contribution, α_r, from a rigid rotation of the surface (plus any built-in twist, although none is assumed to exist here). The second component is the elastic angle of twist $\theta(y)$. Hence

$$\alpha(y) = \alpha_r + \theta(y) \qquad (4.40)$$

where, as is appropriate for strip theory, the contribution from downwash associated with vortices at the wing tip is neglected. Therefore, associated with the angle of attack at each infinitesimal section is a component of sectional-lift coefficient given by strip theory as

$$c_\ell(y) = a[\alpha_r + \theta(y)] \qquad (4.41)$$

4.2.2 Equilibrium Equation

Because we are analyzing the static behavior of this wing, it is appropriate to simplify the fundamental constitutive relationship of torsional deformation, Eq. (2.42), to read

$$T = \overline{GJ}\frac{d\theta}{dy} \qquad (4.42)$$

where $\overline{GJ}$ is the effective torsional stiffness and T is the twisting moment about the elastic axis. Now, a static equation of moment equilibrium about the elastic axis can be obtained by equating the rate of change of twisting moment to the negative of the applied torque distribution. This is a specialization of Eq. (2.43) in which time-dependent terms are ignored, yielding

$$\frac{dT}{dy} = \frac{d}{dy}\left(\overline{GJ}\frac{d\theta}{dy}\right) = -M' \tag{4.43}$$

Recognizing that uniformity implies $\overline{GJ}$ is constant over the length; substituting Eqs. (4.37) into Eq. (4.35) to obtain the applied torque; and, finally, substituting the applied torque and Eq. (4.42) for the internal torque into the equilibrium equation, Eq. (4.43), we obtain

$$\overline{GJ}\frac{d^2\theta}{dy^2} = -qc^2 c_{mac} - eqcc_\ell + Nmgd \tag{4.44}$$

Eq. (4.41) now can be substituted into the equilibrium equation to yield an inhomogeneous, second-order, ordinary differential equation with constant coefficients

$$\frac{d^2\theta}{dy^2} + \frac{qcae}{\overline{GJ}}\theta = -\frac{1}{\overline{GJ}}\left(qc^2 c_{mac} + qcae\alpha_r - Nmgd\right) \tag{4.45}$$

A complete description of this equilibrium condition requires specification of the boundary conditions. Because the surface is built in at the root and free at the tip, these conditions can be written as

$$
\begin{aligned}
y = 0: & \quad \theta = 0 \quad \text{(zero deflection)} \\
y = \ell: & \quad \frac{d\theta}{dy} = 0 \quad \text{(zero twisting moment)}
\end{aligned}
\tag{4.46}
$$

Obviously, these boundary conditions are valid only for the clamped-free condition. The boundary conditions for other end conditions for beams in torsion are given in Section 3.2.2.

4.2.3 Torsional Divergence

If it is presumed that the configuration parameters of the uniform wing are known, then it is possible to solve Eq. (4.45) to determine the resulting twist distribution and associated airload. To simplify the notation, let

$$
\begin{aligned}
\lambda^2 &\equiv \frac{qcae}{\overline{GJ}} \\
\lambda^2\overline{\alpha}_r &\equiv \frac{1}{\overline{GJ}}\left(qc^2 c_{mac} - Nmgd\right)
\end{aligned}
\tag{4.47}
$$

so that

$$\overline{\alpha}_r = \frac{cc_{mac}}{ae} - \frac{Nmgd}{qcae} \tag{4.48}$$

Note that λ^2 and $\overline{\alpha}_r$ are independent of y because the wing is assumed to be uniform. The static-aeroelastic equilibrium equation now can be written as

$$\frac{d^2\theta}{dy^2} + \lambda^2\theta = -\lambda^2(\alpha_r + \overline{\alpha}_r) \tag{4.49}$$

The general solution to this linear ordinary differential equation is

$$\theta = A\sin(\lambda y) + B\cos(\lambda y) - (\alpha_r + \overline{\alpha}_r) \tag{4.50}$$

subject to the condition that $\lambda \neq 0$. Applying the boundary conditions, we find that

$$\begin{aligned} \theta(0) &= 0: \qquad B = \alpha_r + \overline{\alpha}_r \\ \theta'(\ell) &= 0: \qquad A = B\tan(\lambda\ell) \end{aligned} \tag{4.51}$$

where $(\ )' = d(\ )/dy$. Thus, the elastic-twist distribution becomes

$$\theta = (\alpha_r + \overline{\alpha}_r)\left[\tan(\lambda\ell)\sin(\lambda y) + \cos(\lambda y) - 1\right] \tag{4.52}$$

Because θ is now known, the spanwise-lift distribution can be found using the relationship

$$L' = qca(\alpha_r + \theta) \tag{4.53}$$

It is important to note from the expression for elastic twist that θ becomes infinite as $\lambda\ell$ approaches $\pi/2$. This phenomenon is called "torsional divergence" and depends on the numerical value of

$$\lambda = \sqrt{\frac{qcae}{GJ}} \tag{4.54}$$

Thus, it is apparent that there exists a value of the dynamic pressure $q = q_D$, at which $\lambda\ell$ equals $\pi/2$, where the elastic twist theoretically becomes infinite. The value q_D is called the "divergence dynamic pressure" and is given by

$$q_D = \frac{GJ}{eca}\left(\frac{\pi}{2\ell}\right)^2 \tag{4.55}$$

Noting now that we can write

$$\lambda\ell = \frac{\pi}{2}\sqrt{\overline{q}} \tag{4.56}$$

with

$$\overline{q} = \frac{q}{q_D} \tag{4.57}$$

the twist angle of the wing at the tip can be written as

$$\begin{aligned} \theta(\ell) &= (\alpha_r + \overline{\alpha}_r)\left[\sec(\lambda\ell) - 1\right] \\ &= (\alpha_r + \overline{\alpha}_r)\left[\sec\left(\frac{\pi}{2}\sqrt{\overline{q}}\right) - 1\right] \end{aligned} \tag{4.58}$$

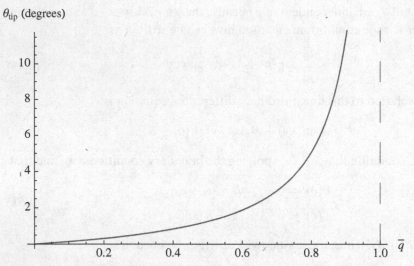

Figure 4.13. Plot of twist angle for the wing tip versus $\overline{q}$ for $\alpha_r + \overline{\alpha}_r = 1°$

where Eq. (4.48) now can be written as

$$\overline{\alpha}_r = \frac{cc_{mac}}{ae} - \frac{4\ell^2 Nmgd}{GJ\pi^2\overline{q}} \tag{4.59}$$

Letting d be zero so that $\overline{\alpha}_r$ becomes independent of $\overline{q}$, we can examine the behavior of $\theta(\ell)$ versus $\overline{q}$. Such a function is plotted in Fig. 4.13, where we see that the tip-twist angle goes to infinity as $\overline{q}$ approaches unity. Note that the character of the plot in Fig. 4.13 is similar to the prebuckling behavior of columns that have imperfections. It is of practical interest to note that the tip-twist angle may become sufficiently large to warrant concern about the structural integrity for dynamic pressures well below q_D. In practice, designers normally require the divergence dynamic pressure to be outside of the vehicle's flight envelope—perhaps by specifying an appropriate factor of safety.

Because this instability occurs at a dynamic pressure that is independent of the right-hand side of Eq. (4.49), as long as the right-hand side is nonzero, it seems possible that the divergence condition could be obtained from the homogeneous equilibrium equation

$$\frac{d^2\theta}{dy^2} + \lambda^2\theta = 0 \tag{4.60}$$

The general solution to this eigenvalue problem of the Sturm–Liouville type is

$$\theta = A\sin(\lambda y) + B\cos(\lambda y) \tag{4.61}$$

for $\lambda \neq 0$. Applying the boundary conditions, we obtain

$$\begin{aligned} \theta(0) = 0: & \quad B = 0 \\ \theta'(\ell) = 0: & \quad A\lambda\cos(\lambda\ell) = 0 \end{aligned} \tag{4.62}$$

If $A = 0$ in the last condition, there is no deflection; this is a so-called trivial solution. Because $\lambda \neq 0$, a nontrivial solution is obtained when $\cos(\lambda \ell) = 0$. This is the "characteristic equation" with solutions given by

$$\lambda_n \ell = (2n - 1)\frac{\pi}{2} \quad (n = 1, 2, \ldots) \tag{4.63}$$

These values are called "eigenvalues." This set of values for $\lambda_n \ell$ corresponds to a set of dynamic pressures

$$q_n = (2n - 1)^2 \left(\frac{\pi}{2\ell}\right)^2 \frac{\overline{GJ}}{eca} \quad (n = 1, 2, \ldots) \tag{4.64}$$

The lowest of these values, q_1, is equal to the divergence dynamic pressure, q_D, previously obtained from the inhomogeneous equilibrium equation. This result implies that there are nontrivial solutions of the homogeneous equation for the elastic twist. In other words, even for cases in which the right-hand side of Eq. (4.49) is zero (i.e., when $\alpha_r + \overline{\alpha}_r = 0$), there is a nontrivial solution

$$\theta_n = A_n \sin(\lambda_n y) \tag{4.65}$$

for each of these discrete values of dynamic pressure. Because A_n is undetermined, the amplitude of θ_n is arbitrary, which means that the effective torsional stiffness is zero whenever the dynamic pressure $q = q_n$. The mode shape θ_1 is the divergence mode shape, which must not be confused with the twist distribution obtained from the inhomogeneous equation.

If the elastic axis is upstream of the aerodynamic center, then $e < 0$ and λ is imaginary in the preceding analysis. The characteristic equation for the divergence condition becomes $\cosh(|\lambda|\ell) = 0$. Because there is no real value of λ that satisfies this equation, the divergence phenomenon does not occur in this case.

4.2.4 Airload Distribution

It has been observed that the spanwise-lift distribution can be determined as

$$L' = qca(\alpha_r + \theta) \tag{4.66}$$

where we recall from Eq. (4.52) that

$$\theta = (\alpha_r + \overline{\alpha}_r)\left[\tan(\lambda \ell)\sin(\lambda y) + \cos(\lambda y) - 1\right] \tag{4.67}$$

and where $\overline{\alpha}_r$ is given in Eq. (4.48). If the lifting surface is a wind-tunnel model of a wing and is fastened to the wind-tunnel wall, then the load factor, N, is equal to unity and α_r can be specified. The resulting computation of L' is straightforward.

If, however, the lifting surface represents half the wing surface of a flying vehicle, the computation of L' is not as direct. Note that the constant $\overline{\alpha}_r$ is a function of N. Thus, for a given value of α_r, there is a corresponding distribution of elastic twist and a particular airload distribution. This airload can be integrated over the vehicle to obtain the total lift, L. Recall that $N = L/W$, where W is the vehicle weight. It is thus apparent that the load factor, N, is related to the rigid angle of attack, α_r, through the elastic twist angle, θ. For this reason, either of the two variables α_r and N can be specified; the other then can be obtained from the total lift L. Assuming a two-winged vehicle with all the lift being generated from the wings, we find

$$L = 2\int_0^\ell L'\, dy \tag{4.68}$$

Substituting for L' and α_r as given herein yields

$$L = 2qca\int_0^\ell \left\{\alpha_r + (\alpha_r + \overline{\alpha}_r)\left[\tan(\lambda\ell)\sin(\lambda y) + \cos(\lambda y) - 1\right]\right\} dy$$

$$= 2qca\ell\left\{(\alpha_r + \overline{\alpha}_r)\left[\frac{\tan(\lambda\ell)}{\lambda\ell}\right] - \overline{\alpha}_r\right\} \tag{4.69}$$

Because $N = L/W$, this expression can be divided by the vehicle weight to yield a relationship for N in terms of α_r and $\overline{\alpha}_r$. This relationship then can be solved simultaneously with the preceding expression for $\overline{\alpha}_r$, Eq. (4.48), in terms of α_r and N. In this manner, $\overline{\alpha}_r$ can be eliminated, providing either a relationship that expresses N in terms of α_r, given by

$$N = \frac{2\overline{GJ}(\lambda\ell)^2\left\{ae\alpha_r + cc_{mac}\left[1 - \frac{\lambda\ell}{\tan(\lambda\ell)}\right]\right\}}{ae\ell\left\{\frac{We\lambda\ell}{\tan(\lambda\ell)} + 2mg\ell d\left[1 - \frac{\lambda\ell}{\tan(\lambda\ell)}\right]\right\}} \tag{4.70}$$

or a relationship that expresses α_r in terms of N

$$\alpha_r = \frac{NW\ell e}{2\overline{GJ}\lambda\ell\tan(\lambda\ell)} + \left[1 - \frac{\lambda\ell}{\tan(\lambda\ell)}\right]\left[\frac{Nmg\ell^2 d}{\overline{GJ}(\lambda\ell)^2} - \frac{cc_{mac}}{ae}\right] \tag{4.71}$$

These relationships permit us to specify a constant α_r and find $N(q)$ or, alternatively, to specify a constant N and find $\alpha_r(q)$. We find that $N(q)$ starts out at zero for $q = 0$. Conversely, $\alpha_r(q)$ starts out at infinity for $q = 0$. The limiting values as $q \to q_D$ depend on the other parameters. These equations can be used to find the torsional deformation and the resulting airload distribution for a specified flight condition.

The calculation of the spanwise aeroelastic airload distribution is immensely practical and is used in industry in two separate ways. First, it is used to satisfy a requirement of aerodynamicists or performance engineers who need to know the total force and moment on the flight vehicle as a function of altitude and flight condition. In this instance, the dynamic pressure q (and altitude or Mach number) and α_r are specified, and the load factor N or total lift L is computed using Eq. (4.70).

A second requirement is that of structural engineers, who must ensure the structural integrity of the lifting surface for a specified load factor N and flight condition. Such a specification normally is described by what is called a V-N diagram. For the conditions of given load factor and flight condition, it is necessary for structural engineers to know the airload distribution to conduct a subsequent loads and stress analysis. When q (and altitude or Mach number) and N are specified, α_r is then determined from Eq. (4.71). Knowing q, α_r, and N, we then use Eq. (4.48) to find $\overline{\alpha}_r$. The torsional deformation, θ, then follows from Eq. (4.67) and the spanwise-airload

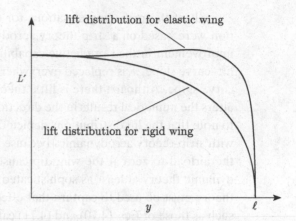

Figure 4.14. Rigid and elastic wing-lift distributions holding α_r constant

distribution follows from Eq. (4.66). From this, the distributions of torsional and bending moments along the wing can be found, leading directly to the maximum stress in the wing, generally somewhere in the root cross section.

Observe that the overall effect of torsional flexibility on the unswept lifting surface is to significantly change the spanwise-airload distribution. This effect can be seen as the presence of the elastic part of the lift coefficient, which is proportional to $\theta(y)$. Because this elastic torsional rotation generally increases as the distance from the root (i.e., out along the span), so also does the resultant airload distribution. The net effect depends on whether α_r or N is specified. If α_r is specified, as in the case of a wall-mounted elastic wind-tunnel model ($N = 1$) or as in performance computations, then the total lift increases with the additional load appearing in the outboard region, as shown in Fig. 4.14.

In the other case, when N is specified by a structural engineer, the total lift (i.e., area under L' versus y) is unchanged, as shown in Fig. 4.15. The addition of lift in the outboard region must be balanced by a decrease inboard. This is accomplished by decreasing α_r as the surface is made more flexible.

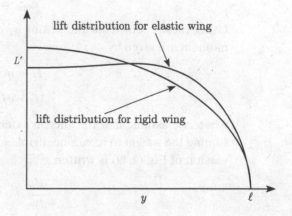

Figure 4.15. Rigid and elastic wing-lift distributions holding total lift constant

All of the preceding equations for torsional divergence and airload distribution were based on a strip-theory aerodynamic representation. A slight numerical improvement in their predictive capability can be obtained if the two-dimensional lift–curve slope, a, is replaced everywhere by the total (i.e., three-dimensional) lift–curve slope. Although there is little theoretical justification for this modification, it alters the numerical results in the direction of the exact answer. Also, it is important to note that the lift distributions depicted in Figs. 4.14 and 4.15 cannot be generated with strip-theory aerodynamics because strip theory fails to pick up the dropoff of the airload to zero at the wing tip caused by three-dimensional effects. An aerodynamic theory at least as sophisticated as Prandtl's three-dimensional lifting-line theory must be used to capture that effect. In such a case, closed-form expressions such as those of Eqs. (4.70) and (4.71) cannot be obtained; instead, it is necessary to use numerical methods to find N as a function of α_r or α_r as a function of N.

4.2.5 Aileron Reversal

In Section 4.1.4, an example illustrating aileron reversal is presented based on a rigid, two-dimensional wing with a flexible support. In this section, we examine the same physical phenomenon using a torsionally flexible wing model. With the geometry and boundary conditions of the uniform, torsionally flexible lifting surface as before, we can derive the reversal dynamic pressure for a clamped-free wing. Two logical choices are presented regarding the defining condition. One is to define reversal dynamic pressure as that dynamic pressure at which the change of total lift with respect to the aileron deflection is equal to zero. Another equally valid definition is to define it as the dynamic pressure at which the change in root-bending moment with respect to the aileron deflection is equal to zero. Finally, we look at the effectiveness of ailerons for roll control—often termed the "roll effectiveness"—of a simplified flying aircraft model.

Note that the presence of an aileron requires that we modify the sectional lift and pitching moment coefficients, so that

$$c_\ell = a\alpha + c_{\ell_\beta}\beta$$
$$c_{mac} = c_{m_\beta}\beta \tag{4.72}$$

Using these coefficients and setting α_r equal to zero, the sectional lift and pitching moment are given by

$$L' = qc\left(a\theta + c_{\ell_\beta}\beta\right)$$
$$M' = eL' + qc^2 c_{m_\beta}\beta \tag{4.73}$$

where we assume that the aileron extends along the entire length of the wing. Assuming the weight to have a negligible effect on the reversal condition, the modified version of Eq. (4.49) is written as

$$\frac{d^2\theta}{dy^2} + \lambda^2\theta = -\lambda^2\psi\beta \tag{4.74}$$

where

$$\psi = \frac{e\,c_{\ell_\beta} + c\,c_{m_\beta}}{a\,e} \tag{4.75}$$

and, as before

$$\lambda^2 \equiv \frac{qcae}{GJ} \tag{4.76}$$

With the boundary conditions of a clamped-free beam, the solution for θ is given by

$$\theta = -\psi\beta\left[1 - \cos(\lambda y) - \sin(\lambda y)\tan(\lambda\ell)\right] \tag{4.77}$$

The total lift for the uniform lifting surface then is obtained as

$$L = \int_0^\ell L'\,dy = \frac{qc\ell}{e}\left[\left(e\,c_{\ell_\beta} + c\,c_{m_\beta}\right)\frac{\tan(\lambda\ell)}{\lambda\ell} - c\,c_{m_\beta}\right]\beta \tag{4.78}$$

Therefore, from the first definition of aileron reversal

$$\frac{\partial L}{\partial\beta} = 0 \Rightarrow \frac{\tan(\lambda\ell)}{\lambda\ell} = \frac{c\,c_{m_\beta}}{c\,c_{m_\beta} + e\,c_{\ell_\beta}} \tag{4.79}$$

which, given e/c and the sectional coefficients,[1] may be solved numerically for $\lambda\ell$. The smallest value of $\lambda\ell$ denoted by $\lambda_1\ell$ yields the reversal dynamic pressure as

$$q_R = \frac{(\lambda_1\ell)^2 \overline{GJ}}{eca\ell^2} \tag{4.80}$$

We may refine the theoretical result by considering a simplified correction from three-dimensional effects by use of a tip-loss factor, typically chosen as $B = 0.97$. Instead of obtaining the total lift by integrating the sectional lift over the entire wing length from $y = 0$ to $y = \ell$, we integrate only from $y = 0$ to $y = B\ell$.

Similarly, we may account for an aileron that does not extend over the entire length of the wing. Suppose that the aileron starts at $y = r\ell$ and extends to $y = R\ell$ with $0 \le r < R \le 1$. This means that there are as many as three segments to be analyzed. There is no inhomogeneous term for the segments between $y = r\ell$ and $y = R\ell$, so instead of Eq. (4.74), we write

$$\frac{d^2\theta_1}{dy^2} + \lambda^2\theta_1 = 0 \qquad 0 \le y \le r\ell$$

$$\frac{d^2\theta_2}{dy^2} + \lambda^2\theta_2 = -\lambda^2\psi\beta \qquad r\ell \le y \le R\ell \tag{4.81}$$

$$\frac{d^2\theta_3}{dy^2} + \lambda^2\theta_3 = 0 \qquad R\ell \le y \le \ell$$

---

[1] Estimated values of the airfoil coefficients may be obtained from experiment or from XFOIL, a computer code based on a panel method for design and analysis of subsonic isolated airfoils (see Drela, 1992).

and obtain the resulting six arbitrary constants by imposing the six boundary conditions

$$\theta_1(0) = 0$$

$$\theta_1(r\ell) = \theta_2(r\ell)$$

$$\frac{d\theta_1}{dy}(r\ell) = \frac{d\theta_2}{dy}(r\ell)$$

$$\theta_2(R\ell) = \theta_3(R\ell)$$ (4.82)

$$\frac{d\theta_2}{dy}(R\ell) = \frac{d\theta_3}{dy}(R\ell)$$

$$\frac{d\theta_3}{dy}(\ell) = 0$$

Calculation of the reversal dynamic pressure from the second definition (i.e., the one in terms of the root-bending-moment criterion) is left as an exercise for readers (see Problem 20).

This treatment can be generalized easily to consider the roll effectiveness of a complete aircraft model. Similar problems can be posed in the framework of dynamics, in which the objective is, say, to predict the angular acceleration caused by deflection of a control surface, or the time to change the orientation of the aircraft from one roll angle to another. Depending on the aircraft and the maneuver, it may be necessary to consider nonlinearities. Here, however, only a static, linear treatment is included.

Consider a rolling aircraft with unswept wings, the right half of which is shown in Fig. 4.16, with a constant roll rate denoted by p. As shown in Fig. 4.17, the wing section has an incidence angle with respect to the freestream velocity of $\alpha_r + \theta(y)$. In a roll maneuver with $p > 0$, the right wing moves upward while the left wing moves downward. The right wing then "sees" an additional component of wind velocity equal to py perpendicular to the freestream velocity and downward. As shown in Fig. 4.17, because $py \ll U$, the angle of attack is reduced from the incidence angle to $\alpha_r + \theta - py/U$.

Some contributions to the lift and pitching moment are the same (opposite) on both sides of the aircraft; these are referred to as symmetric (antisymmetric) components. Separate problems can be posed in terms of symmetric and antisymmetric parts, which are generally uncoupled from one another. In particular, we can treat the roll problem as an antisymmetric problem noting that all symmetric components cancel out in pure roll. Hence, we can discard them a priori. For example, in the relationship

$$\alpha = \overset{0}{\cancel{\alpha_r}} + \theta(y) - \frac{py}{U}$$ (4.83)

the first term, α_r, drops out because of symmetry. Both $\theta(y)$ and the roll-rate term are antisymmetric because θ and β have the opposite sense across the mid-plane of the aircraft. The last term, which represents the increment in the angle of attack from

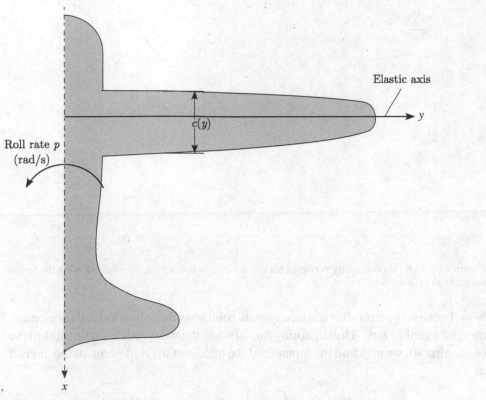

Figure 4.16. Schematic of a rolling aircraft

the roll rate p based on the assumption of a small angle of attack, also is explicitly antisymmetric.

Assuming $c(y)$ to be a constant, c, we may write the governing differential equation as

$$\frac{d^2\theta}{dy^2} + \lambda^2\theta = \lambda^2\left(\frac{py}{U} - \beta\psi\right) \tag{4.84}$$

with boundary conditions $\theta(0) = d\theta/dy(\ell) = 0$. The solution, with terms removed that are symmetric about $y = 0$, is

$$\theta = \frac{p}{U\lambda}\left[\lambda y - \sec(\lambda\ell)\sin(\lambda y)\right] + \psi\beta\tan(\lambda\ell)\sin(\lambda y) \tag{4.85}$$

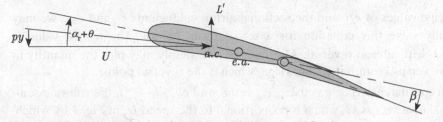

Figure 4.17. Section of right wing with positive aileron deflection

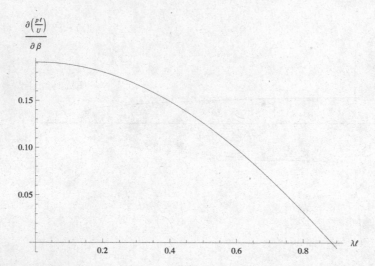

Figure 4.18: Roll rate sensitivity versus $\lambda\ell$ for $e = 0.25c$, $c_{\ell_\beta} = 0.8$, and $c_{m_\beta} = -0.5$, showing the reversal point at $\lambda\ell = 0.885285$

Now, because the aircraft is in a steady-state rolling motion, the total rolling moment must be equal to zero. Thus, ignoring the offset of the wing root from the mid-plane of the aircraft, we may find the moment of the lift about the mid-plane of the aircraft as

$$\int_0^\ell yL'(y)dy = \int_0^\ell yqc\left[a\left(\theta - \frac{py}{U}\right) + c_{\ell_\beta}\beta\right]dy \tag{4.86}$$

which, when set equal to zero, can be solved for the constant roll rate p. (Note that the three terms in Eq. [4.86] are the contributions toward the rolling moment due to the elastic twist, the roll rate p, and the aileron deflection β, respectively.) This result, written here in dimensionless form as $p\ell/U$, is given by

$$\frac{p\ell}{U} = \frac{\lambda\ell\beta\left(ec_{\ell_\beta}\{(\lambda\ell)^2 + 2\sin(\lambda\ell)[\tan(\lambda\ell) - \lambda\ell]\} + 2cc_{m_\beta}\sin(\lambda\ell)[\tan(\lambda\ell) - \lambda\ell]\right)}{2ae[\tan(\lambda\ell) - \lambda\ell]} \tag{4.87}$$

which is proportional to β. At a certain dynamic pressure, we are unable to change the roll rate by changing β. This dynamic pressure occurs when the sensitivity of the roll rate to β vanishes; viz.

$$\frac{\partial\left(\frac{p\ell}{U}\right)}{\partial\beta} = \frac{\lambda\ell\left(ec_{\ell_\beta}\{(\lambda\ell)^2 + 2\sin(\lambda\ell)[\tan(\lambda\ell) - \lambda\ell]\} + 2cc_{m_\beta}\sin(\lambda\ell)[\tan(\lambda\ell) - \lambda\ell]\right)}{2ae[\tan(\lambda\ell) - \lambda\ell]} = 0 \tag{4.88}$$

For specific values of e/c and the sectional airfoil coefficients c_{ℓ_β} and c_{m_β}, we may numerically solve this equation for a set of roots for $\lambda\ell$. The lowest value is associated with aileron reversal. Alternatively, we simply may plot the quantity in Eq. (4.88) versus $\lambda\ell$ until it changes sign, which is the reversal point.

For a specific case (i.e., $e = 0.25c$, $c_{\ell_\beta} = 0.8$, and $c_{m_\beta} = -0.5$), the roll-rate sensitivity is shown versus $\lambda\ell$, which is proportional to the speed U, in Fig. 4.18, which shows the reversal point at $\lambda\ell = 0.984774$. Notice that the curve at low speed starts

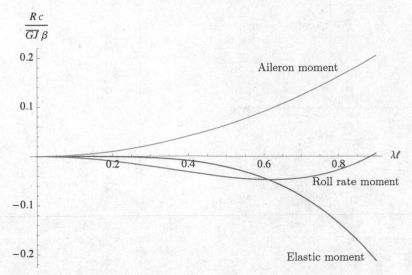

Figure 4.19. Contributions to rolling moment R (normalized) from the three terms of Eq. (4.86)

as relatively flat and monotonically decreases until the reversal point is reached. This shape is a typical result and shows the importance of static aeroelasticity in this aspect of flight mechanics. It is also interesting to observe the relative contributions to the rolling moment from the elastic twist, the rolling motion, and the aileron deflections depicted in Fig. 4.19. At the reversal point, p vanishes, and the rolling moment contributions from elastic twist and from aileron deflection exactly cancel out one another.

4.2.6 Sweep Effects

To observe the effect of sweeping a wing aft or forward on the aeroelastic characteristics, it is presumed that the swept geometry is obtained by rotating the surface about the root of the elastic axis, as illustrated in Fig. 4.20. The aerodynamic reactions depend on the angle of attack as measured in the streamwise direction as

$$\alpha = \alpha_r + \theta \tag{4.89}$$

where θ is the change in the streamwise angle of attack caused by elastic deformation. To develop a kinematical relationship for θ, we introduce the unit vectors $\hat{\mathbf{a}}_1$ and $\hat{\mathbf{a}}_2$, aligned with the y axis and the freestream, respectively. Another set of unit vectors, $\hat{\mathbf{b}}_1$ and $\hat{\mathbf{b}}_2$, is obtained by rotating $\hat{\mathbf{a}}_1$ and $\hat{\mathbf{a}}_2$ by the sweep angle Λ, as shown in Fig. 4.20, so that $\hat{\mathbf{b}}_1$ is aligned with the elastic axis (i.e., the $\bar{y}$ axis). From Fig. 4.20, we see that

$$\hat{\mathbf{b}}_1 = \cos(\Lambda)\hat{\mathbf{a}}_1 + \sin(\Lambda)\hat{\mathbf{a}}_2$$
$$\hat{\mathbf{b}}_2 = -\sin(\Lambda)\hat{\mathbf{a}}_1 + \cos(\Lambda)\hat{\mathbf{a}}_2 \tag{4.90}$$

Observe that the total rotation of the local wing cross-sectional frame caused by elastic deformation can be written as the combination of rotations caused by wing

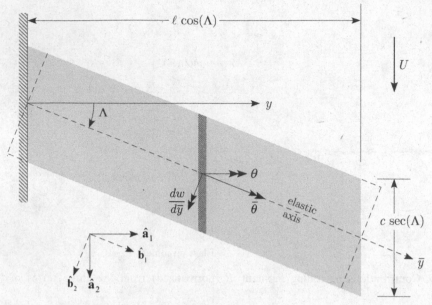

Figure 4.20. Schematic of swept wing (positive Λ)

torsion, $\overline{\theta}$ about $\hat{\mathbf{b}}_1$, and wing bending, $dw/d\overline{y}$ about $\hat{\mathbf{b}}_2$, where w is the bending deflection (positive up, which in Fig. 4.20 is out of the paper). Now, θ is the component of this total rotation about $\hat{\mathbf{a}}_1$; that is

$$
\begin{aligned}
\theta &= \left(\overline{\theta} \hat{\mathbf{b}}_1 + \frac{dw}{d\overline{y}} \hat{\mathbf{b}}_2 \right) \cdot \hat{\mathbf{a}}_1 \\
&= \overline{\theta} \cos(\Lambda) - \frac{dw}{d\overline{y}} \sin(\Lambda)
\end{aligned}
\tag{4.91}
$$

From this relationship, it can be noted that as the result of sweep, the effective angle of attack is altered by bending. This coupling between bending and torsion affects both the static aeroelastic response of the wing in flight as well as the conditions under which divergence occurs. Also, it can be observed that for combined bending and torsion of a swept, elastic wing, the section in the direction of the streamwise airflow exhibits a change in camber—a higher-order effect that is here neglected.

To facilitate direct comparison with the previous unswept results, to the extent possible, the same structural and aerodynamic notation is retained as was used for the unswept planform. To determine the total elastic deflection, two equilibrium equations are required: one for torsional moment equilibrium as in the unswept case and one for transverse force equilibrium (associated with bending). These equations can be written as

$$
\begin{aligned}
\frac{d}{d\overline{y}} \left(\overline{GJ} \frac{d\overline{\theta}}{d\overline{y}} \right) &= -qec\overline{a}\alpha - qc^2\overline{c}_{mac} + Nmgd \\
\frac{d^2}{d\overline{y}^2} \left(\overline{EI} \frac{d^2 w}{d\overline{y}^2} \right) &= qc\overline{a}\alpha - Nmg
\end{aligned}
\tag{4.92}
$$

In these equilibrium equations, $\bar{a}$ is used to denote the two-dimensional lift–curve slope of the swept surface and $\bar{c}_{mac}$ to represent the two-dimensional pitching-moment coefficient of the swept surface. These aerodynamic constants are related to their unswept counterparts by

$$\bar{a} = a \cos(\Lambda)$$
$$\bar{c}_{mac} = c_{mac} \cos^2(\Lambda) \tag{4.93}$$

for moderate- to high-aspect-ratio surfaces. Substituting for $\bar{a}$, $\alpha = \alpha_r + \theta$ and, in turn, the dependence of θ on $\bar{\theta}$ and w from Eq. (4.91), specializing for spanwise uniformity so that $\overline{GJ}$ and $\overline{EI}$ are constants, and letting $(\ )'$ denote $d(\ )/d\bar{y}$, we obtain two coupled, ordinary differential equations for torsion and bending given by

$$\bar{\theta}'' + \frac{qeca}{\overline{GJ}}\bar{\theta}\cos^2(\Lambda) - \frac{qeca}{\overline{GJ}}w'\sin(\Lambda)\cos(\Lambda)$$

$$= -\frac{1}{\overline{GJ}}\left[qeca\alpha_r\cos(\Lambda) + qc^2 c_{mac}\cos^2(\Lambda) - Nmgd\right]$$

$$w'''' + \frac{qca}{\overline{EI}}w'\sin(\Lambda)\cos(\Lambda) - \frac{qca}{\overline{EI}}\bar{\theta}\cos^2(\Lambda) = \frac{1}{\overline{EI}}\left[qca\alpha_r\cos(\Lambda) - Nmg\right] \tag{4.94}$$

Because the surface is built in at the root and free at the tip, the following boundary conditions must be imposed on the solution:

$$
\begin{array}{llll}
\bar{y} = 0: & \bar{\theta} & = 0 & \text{(zero torsional rotation)} \\
& w & = 0 & \text{(zero deflection)} \\
& w' & = 0 & \text{(zero bending slope)} \\
\bar{y} = \ell: & \bar{\theta}' & = 0 & \text{(zero twisting moment)} \\
& w'' & = 0 & \text{(zero bending moment)} \\
& w''' & = 0 & \text{(zero shear force)}
\end{array}
\tag{4.95}
$$

Bending-torsion coupling is exhibited in Eqs. (4.94) through the term involving w in the torsion equation and through the term involving $\bar{\theta}$ in the bending equation.

There are two special cases of interest in which the coupling either vanishes or is much simplified so that we can solve the equations analytically. The first is for the case of vanishing sweep in which the uncoupled torsion equation (i.e., the first of Eqs. [4.94]) is the same as previously discussed and clearly leads to solutions for either the torsional divergence condition or the torsional deformation and air-load distribution as discussed (see Sections 4.2.3 and 4.2.4, respectively). In the latter case, once the torsional deformation is obtained, the solution for $\theta = \bar{\theta}$ can be substituted into the bending equation (i.e., the second of Eqs. [4.94]). Integration of the resulting ordinary differential equation and application of the boundary conditions lead to the shear force, bending moment, bending slope, and bending deflection.

A second special case occurs when $e = 0$. In this case, torsional divergence does not take place, and a polynomial solution for $\bar{\theta}$ can be found from the $\bar{\theta}$ equation and boundary conditions. Substitution of this solution into the bending equation leads

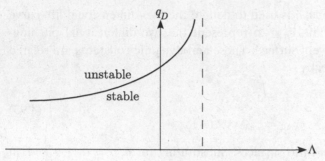

Figure 4.21. Divergence dynamic pressure versus Λ

to a fourth-order, ordinary differential equation for w with a polynomial forcing function; note that the $\bar{\theta}$ terms are now part of that forcing function. This equation and accompanying boundary conditions can be solved for the bending deflection, but the solution is not straightforward. Alternatively, to solve this equation for a divergence condition, we need only the homogeneous part, which can be written as a third-order equation in $\zeta = w'$; namely

$$\zeta''' + \frac{qca}{EI}\zeta \sin(\Lambda)\cos(\Lambda) = 0 \tag{4.96}$$

For the clamped-free boundary conditions $\zeta(0) = \zeta'(\ell) = \zeta''(\ell) = 0$, this equation has a known analytical solution that yields a divergence dynamic pressure of

$$q_D = -6.32970\frac{\overline{EI}}{ac\ell^3 \sin(\Lambda)\cos(\Lambda)} \tag{4.97}$$

The minus sign implies that this bending-divergence instability takes place only for forward-swept wings; that is, where $\Lambda < 0$.

Examination of Eqs. (4.94) illustrates that there are two ways in which the sweep influences the aeroelastic behavior. One way is the loss of aerodynamic effectiveness, as exhibited by the change in the second term of the torsion equations from

$$\frac{qeca}{GJ}\bar{\theta} \qquad \text{to} \qquad \frac{qeca}{GJ}\bar{\theta}\cos^2(\Lambda) \tag{4.98}$$

Note that this effect is independent of the direction of sweep. The second effect is the influence of bending slope on the effective angle of attack (see Eq. 4.91), which leads to bending-torsion coupling. This coupling has a strong influence on both divergence and load distribution. The total effect of sweep depends strongly on whether the surface is swept backward or forward. This can be illustrated by its influence on the divergence dynamic pressure, q_D, as shown in Fig. 4.21. It is apparent that forward sweep causes the surface to be more susceptible to divergence, whereas backward sweep increases the divergence dynamic pressure. Indeed, a small amount of backward sweep (i.e., for the idealized case under consideration, depending on e/ℓ and $\overline{GJ/EI}$, only 5 or 10 degrees) can cause the divergence dynamic pressure to become sufficiently large that it ceases to be an issue. Specific cases are discussed later in this section in conjunction with an approximate solution of the governing equations.

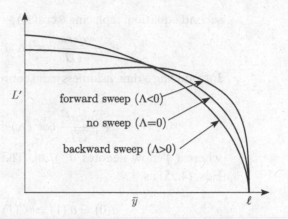

Figure 4.22. Lift distribution for positive, zero, and negative Λ

The overall effect of sweep on the aeroelastic-load distribution also strongly depends on whether the surface is swept forward or backward. This is illustrated in Fig. 4.22, which shows spanwise load distributions for an elastic surface for which the total lift (or N) is held constant by adjusting α_r. From the standpoint of structural loads, it is apparent that the root bending moment is significantly greater for forward sweep than for backward sweep at a given value of total lift.

The primary motivation for sweeping a lifting surface is to improve the vehicle performance through drag reduction, although some loss in lifting capability may be experienced. However, these aeroelastic considerations can have a significant impact on design decisions. From an aeroelastic standpoint, forward sweep exacerbates divergence instability and increases structural loads, whereas backward sweep can alleviate these concerns. The advent of composite lifting surfaces enabled the use of bending-twist elastic coupling to passively stabilize forward sweep, making it possible to use forward-swept wings. Indeed, the X-29 could not have been flown without a means to stabilize the wings against divergence. We discuss this further in Section 4.2.7.

Exact Solution for Bending-Torsion Divergence. Extraction of the analytical solution of the set of coupled, ordinary differential equations in Eqs. (4.94) is complicated. The exact analytical solution is obtained most easily by first converting the coupled set of equations into a single equation governing the elastic component of the angle of attack. For calculation of only the divergence dynamic pressure, we can consider just the homogeneous parts of Eqs. (4.94):

$$\bar{\theta}'' + \frac{qeca}{GJ}\bar{\theta}\cos^2(\Lambda) - \frac{qeca}{GJ}w'\sin(\Lambda)\cos(\Lambda) = 0$$

$$w'''' + \frac{qca}{EI}w'\sin(\Lambda)\cos(\Lambda) - \frac{qca}{EI}\bar{\theta}\cos^2(\Lambda) = 0$$

(4.99)

To obtain a single equation, we differentiate the first equation with respect to $\bar{y}$ and multiply it by $\cos(\Lambda)$. From this modified first equation, we subtract $\sin(\Lambda)$ times the second equation, replacing $\bar{\theta}\cos(\Lambda) - w'\sin(\Lambda)$ with θ, to obtain

$$\theta''' + \frac{qeca}{\overline{GJ}}\cos^2(\Lambda)\theta' + \frac{qca}{EI}\sin(\Lambda)\cos(\Lambda)\theta = 0 \qquad (4.100)$$

Introducing a dimensionless axial coordinate $\eta = \bar{y}/\ell$, this equation can be written as

$$\theta''' + \frac{qeca\ell^2}{\overline{GJ}}\cos^2(\Lambda)\theta' + \frac{qca\ell^3}{EI}\sin(\Lambda)\cos(\Lambda)\theta = 0 \qquad (4.101)$$

where $(\;)'$ now denotes $d(\;)/d\eta$. The boundary conditions can be derived from Eqs. (4.95) as

$$\theta(0) = \theta'(1) = \theta''(1) + \frac{qeca\ell^2}{\overline{GJ}}\cos^2(\Lambda)\theta(1) = 0 \qquad (4.102)$$

Here, the first of Eqs. (4.99) and the final boundary condition from Eqs. (4.95) are used to derive the third boundary condition.

The exact solution for Eqs. (4.101) and (4.102) was obtained by Diederich and Budiansky (1948). Its behavior is complex, with multiple branches, and it is not used easily in a design context. However, a simple approximation of one branch is presented next and compared with plots of the exact solution.

Approximate Solution for Bending-Torsion Divergence. In view of the complexity of the exact solution, it is fortunate that there are various approximate methods for treating such equations, one of which is the application of the Ritz method to the principle of virtual work (see Section 3.5). In this special case, the kinetic energy is zero, and the resulting algebraic equations are a special case of the generalized equations of motion (see Section 3.1.5), termed "generalized equations of equilibrium." Determination of such an approximate solution is left as an exercise for readers (see Problems 11–16).

Here, we consider instead an approximation of one branch of the analytical solution for the bending-torsion divergence problem. Fortunately, the most important branch from a physical point of view behaves simply. Indeed, if we define

$$\tau = \frac{qeca\ell^2}{\overline{GJ}}\cos^2(\Lambda)$$

$$\beta = \frac{qca\ell^3}{EI}\sin(\Lambda)\cos(\Lambda) \qquad (4.103)$$

then, as shown by Diederich and Budiansky (1948), the divergence boundary can be approximately represented within a certain range in terms of a straight line

$$\tau_D = \frac{\pi^2}{4} + \frac{3\pi^2}{76}\beta_D \qquad (4.104)$$

Note that for a wing rigid in bending, we have $\beta_D = 0$ and, thus, $\tau_D = \frac{\pi^2}{4}$, which is the exact solution for pure torsional divergence. Also, for a torsionally rigid wing, we have $\tau_D = 0$ and, thus, $\beta_D = -19/3$, which is very close to -6.3297, the exact solution for bending divergence. For the cases in between, the error is quite small.

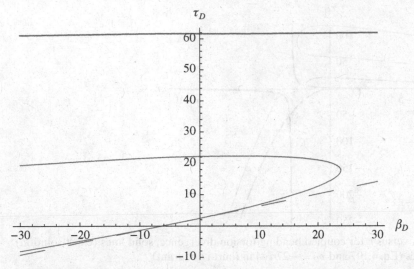

Figure 4.23. τ_D versus β_D for coupled bending-torsion divergence; solid lines (exact solution) and dashed line (Eq. 4.104).

It is important to note that the sign of τ is driven by the sign of e, whereas the sign of β is driven by the sign of Λ. The approximate solution in Eq. (4.104) is plotted along with some branches of the exact solution in Fig. 4.23. Note the excellent agreement between the straight-line approximation and the exact solution near the origin. Note also that the intersections of the solution with the τ_D axis (where $\beta_D = 0$) coincide with the squares of the roots previously obtained in Section 4.2.3, Eq. (4.63), as $(2n-1)^2\pi^2/4$ for $n = 1, 2, \ldots, \infty$ (i.e., $\pi^2/4, 9\pi^2/4, \ldots$).

A more convenient way of depicting the behavior of the divergence dynamic pressure is to plot τ_D versus a parameter that depends on only the configuration. This can be accomplished by introducing the dimensionless parameter r, given by

$$r \equiv \frac{\beta}{\tau} = \frac{\ell}{e}\frac{\overline{GJ}}{\overline{EI}}\tan(\Lambda) \tag{4.105}$$

which can be positive, negative, or zero. Equation (4.104) can then be written as

$$\tau_D = \frac{\pi^2}{4} + \frac{3\pi^2 r}{76}\tau_D \tag{4.106}$$

Thus, we can solve for τ_D such that

$$\tau_D = \frac{\pi^2}{4\left(1 - \frac{3\pi^2 r}{76}\right)} \tag{4.107}$$

or alternatively for q_D, equal to

$$q_D = \frac{\overline{GJ}\pi^2}{4eca\ell^2\cos^2(\Lambda)\left[1 - \frac{3\pi^2}{76}\frac{\ell}{e}\frac{\overline{GJ}}{\overline{EI}}\tan(\Lambda)\right]} \tag{4.108}$$

Several branches of the exact solution of Eqs. (4.101) and (4.102) for the smallest absolute values of τ_D versus r are plotted as solid lines in Fig. 4.24. Note that there is

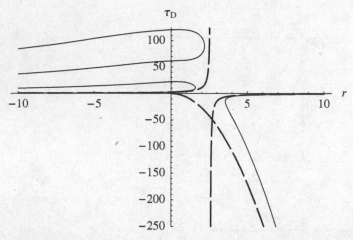

Figure 4.24. τ_D versus r for coupled bending-torsion divergence; solid lines (exact solution) and dashed lines (Eq. 4.107 and $\tau_D = -27r^2/4$ in fourth quadrant)

at least one branch in all quadrants except the third, and there is only one branch in the fourth quadrant. The approximate solutions for τ_D versus r from Eq. (4.107) are plotted as dashed hyperbolae in the first, second, and fourth quadrants. Moreover, as r becomes large, the solution in the fourth quadrant asymptotically approaches the parabola $\tau_D = -27r^2/4$, also shown as a dashed curve. Note that as in Fig. 4.23, the intersections of the roots with the τ_D axis are $\pi^2/4, 9\pi^2/4, 25\pi^2/4$, and so on. The configuration of any wing fixes the value of r. For positive e, we consider only positive values of τ_D. Thus, we start from zero and proceed in the positive τ_D direction on this plot (i.e., at constant r) to find the first intersection with a solid line. This value of τ_D is the normalized dynamic pressure at which divergence occurs. In Fig. 4.25,

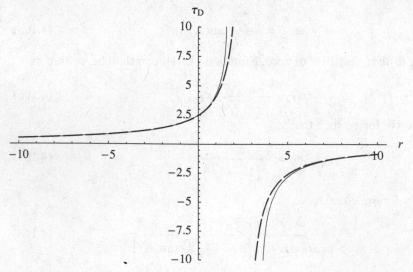

Figure 4.25. τ_D versus r for coupled bending-torsion divergence; solid lines (exact solution) and dashed lines (Eq. 4.107)

an enlargement of these results in a more practical range is shown. It is easily seen that the dashed lines in the first and second quadrants are close to the solid lines when $r < 1.5$. Note that when $e < 0$, a negative value of τ_D leads to a positive value of q_D. In this case, we should proceed along a line of constant r in the negative τ_D direction.

It is interesting that the approximate solution, despite its proximity to the exact solution, exhibits a qualitatively different behavior mathematically. The approximate solution exhibits an asymptotic behavior, with τ_D tending to plus infinity from the left and to minus infinity from the right at the value of r that causes the denominator to vanish—namely, when $r = 76/(3\pi^2) = 2.56680$. If the approximate solution were exact, mathematically it would mean that divergence is not possible at that value of r. Moreover, physically it would mean that divergence is not possible for $e > 0$ and $r \geq 76/(3\pi^2)$ [or for $e < 0$ and $r \leq 76/(3\pi^2)$]. Actually, however, the exact solution exhibits an instability of the "limit-point" variety. For $e > 0$, this means that divergence occurs for small and positive values of r. Moreover, as r is increased in the first quadrant, τ_D also increases until a certain point is reached, at which two things happen: (1) above this value of τ_D, the curve turns back to the left instead of reaching an asymptote; and (2) any slight increase in r beyond this point causes the solution to jump to a higher branch. This point is called a limit point. On the main branch of the curve in the first quadrant, for example, the limit point is at $r = 1.59768$ and $\tau_D = 10.7090$. It is shown in the plot in Fig. 4.24 that any slight increase in r causes the solution to jump from the lower branch—where its value is 10.7090—to a higher branch, where its value is 66.8133, at which point τ_D is rapidly increasing with r. So, although there is no value of r that results in an infinite exact value of the divergence dynamic pressure, practically speaking, divergence in the vicinity of the limit-point value of r is all but eliminated. Thus, it is sufficient for practical purposes to say that divergence is not possible near those points where the approximate solution blows up, and we may regard the approximate solution as sufficiently close to the exact solution for design purposes. The limit point in the fourth quadrant is appropriate for the situation in which $e < 0$—namely, when the aerodynamic center is behind the elastic axis. There, the exact limit point is at $r = 3.56595$ and $\tau_D = -14.8345$. Note that the negative values of e and τ_D yield a positive q_D. It is left to readers as an exercise to explore this possibility further (see Problem 18).

Although there are qualitative differences, as noted, between the exact and approximate solutions, within the practical range of interest, this linear approximation of the divergence boundary in terms of τ_D and β_D is numerically accurate and leads to a simple expression for the divergence dynamic pressure in terms of the structural stiffnesses, e/ℓ, and the sweep angle (i.e., Eq. [4.108]). This approximate formula can be used in design to explore the behavior of the divergence dynamic pressure as a function of the various configuration parameters therein. For the purpose of displaying results for the divergence dynamic pressure when $e > 0$, it is convenient to normalize q_D with its value at zero sweep angle; namely

$$q_{D_0} = \frac{\pi^2 \overline{GJ}}{4eca\ell^2} \tag{4.109}$$

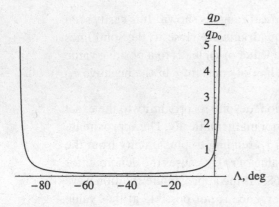

Figure 4.26. Normalized divergence dynamic pressure for an elastically uncoupled, swept wing with $\overline{GJ}/\overline{EI} = 1.0$ and $e/\ell = 0.02$

so that

$$\frac{q_D}{q_{D_0}} = \frac{1 + \tan^2(\Lambda)}{1 - \frac{3\pi^2}{76}\frac{\ell}{e}\frac{\overline{GJ}}{\overline{EI}}\tan(\Lambda)} \tag{4.110}$$

Thus, for a wing structural design with given values of e, $\overline{GJ}$, $\overline{EI}$, and ℓ, there are values of sweep angle Λ for which the divergence dynamic pressure goes to infinity or becomes negative, implying that divergence is not possible at those values of Λ. Some values of Λ make the numerator infinite because $\tan(\Lambda)$ blows up, whereas other values make the denominator vanish or switch signs. Therefore, within the principal range of $-90° \leq \Lambda \leq 90°$, we can surmise that divergence can take place only for cases in which $|\Lambda| \neq 90°$ and $3\pi^2 r \neq 76$. Sign changes have the following consequences: Divergence is possible only if $-90° < \Lambda < \Lambda_\infty$, where

$$\tan(\Lambda_\infty) = \frac{76\overline{EI}e}{3\pi^2\overline{GJ}\ell} \tag{4.111}$$

Thus, Eq. (4.110) can be written as

$$\frac{q_D}{q_{D_0}} = \frac{1 + \tan^2(\Lambda)}{1 - \frac{\tan(\Lambda)}{\tan(\Lambda_\infty)}} \tag{4.112}$$

In other words, we avoid divergence by choosing $\Lambda \geq \Lambda_\infty$, and the divergence dynamic pressure drops drastically as Λ is decreased below Λ_∞. Because Λ_∞ is likely to be small, this frequently means that backswept wings are free of divergence and that divergence dynamic pressure drops drastically for forward-swept wings. Because Λ_∞ is the asymptotic value of Λ from the approximate solution, which is greater than the limit-point value of Λ from the exact solution, we may surmise that the approximate solution provides a conservative design. Figure 4.26 shows the behavior of divergence dynamic pressure for a wing with $\overline{GJ}/\overline{EI} = 1.0$ and $e/\ell = 0.02$. The plot, as expected, passes through unity when the sweep angle is zero. Because Λ_∞ is very small for this case, the divergence dynamic pressure goes to infinity for a very small positive value of sweep angle. Thus, even a small angle of backward sweep can make divergence impossible. Figure 4.27 shows the result of decreasing $\overline{GJ}/\overline{EI}$ to 0.2 and holding e/ℓ constant. Because Λ_∞ increases, the wing must be swept back farther than in the previous case to avoid divergence.

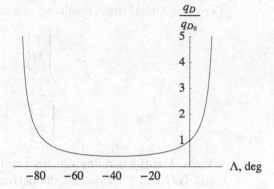

Figure 4.27. Normalized divergence dynamic pressure for an elastically uncoupled, swept wing with $\overline{GJ}/\overline{EI} = 0.2$ and $e/\ell = 0.02$

Because e can be positive, negative, or zero, q_{D_0} also may be positive, negative, or zero. Thus, by normalizing q_D by q_{D_0}, we may obfuscate the role of the sign of e on q_D. In such cases and perhaps others, it is more convenient to write Eq. (4.112) in a form that does not depend on q_{D_0}. One way to accomplish this is to eliminate q_{D_0} from the expression for q_D using Eqs. (4.109) and (4.111), yielding

$$\frac{q_D a c \ell^3}{\overline{EI}} = \frac{19\left[1 + \tan^2(\Lambda)\right]}{3\left[\tan(\Lambda_\infty) - \tan(\Lambda)\right]} \tag{4.113}$$

making it clearer that divergence occurs only when $-90° < \Lambda < \Lambda_\infty$, regardless of the sign of e. This form of the formula also shows more explicitly that $\overline{EI}$ has a role in the design of swept wings that are free of divergence.

4.2.7 Composite Wings and Aeroelastic Tailoring

Aeroelastic tailoring is the design of wings using the directional properties of composite materials to optimize aeroelastic performance. The concept of aeroelastic tailoring is relatively new and came into the forefront during the design of forward-swept wings in the 1980s. Equation (4.112) shows that q_D drops dramatically for forward-swept, untailored wings. The low divergence speed was a major hurdle in the design of wings with forward sweep. As discussed in this section, use of composite materials can help remove the disadvantages of forward sweep. Currently, aeroelastic tailoring is an integral part of the design of composite wings and can be used to improve performance in a variety of ways.

Composite materials are anisotropic, which implies different material characteristics (e.g., stiffness) in different directions. A simple beam model is helpful in developing an understanding of the behavior of composite wings. Such models may exhibit bending-torsion elastic coupling. Analysis of beams with elastic coupling is more involved, but it leads to helpful results.

Let us introduce such coupling in our beam equations. For anisotropic beams with bending-torsion coupling, the "constitutive law" (i.e., the relationship between cross-sectional stress resultants and the generalized strains) changes from

$$\begin{Bmatrix} T \\ M \end{Bmatrix} = \begin{bmatrix} \overline{GJ} & 0 \\ 0 & \overline{EI} \end{bmatrix} \begin{Bmatrix} \overline{\theta}' \\ w'' \end{Bmatrix} \tag{4.114}$$

to

$$\begin{Bmatrix} T \\ M \end{Bmatrix} = \begin{bmatrix} \overline{GJ} & -K \\ -K & \overline{EI} \end{bmatrix} \begin{Bmatrix} \overline{\theta}' \\ w'' \end{Bmatrix} \tag{4.115}$$

where K is the bending-torsion coupling stiffness (with the same dimensions as $\overline{EI}$ and $\overline{GJ}$) and $(\ )'$ indicates the derivative with respect to $\overline{y}$. A positive value of K means that a positive bending deflection will be accompanied by a nose-up twist of the wing, which is destabilizing.

Using the coupled constitutive law, the equations of equilibrium become

$$\left(\overline{GJ}\overline{\theta}' - Kw'' \right)' = -qec\overline{a}\alpha - qc^2\overline{c}_{mac} + Nmgd$$
$$\left(\overline{EI}w'' - K\overline{\theta}' \right)'' = qc\overline{a}\alpha - Nmg \tag{4.116}$$

Consider again a wing that is clamped at the root and free at the tip, so that the boundary conditions that must be imposed on the solution are

$$\begin{aligned}
\overline{y} = 0: \quad & \overline{\theta} = 0 & \text{(zero torsional rotation)} \\
& w = 0 & \text{(zero deflection)} \\
& w' = 0 & \text{(zero bending slope)} \\
\overline{y} = \ell: \quad & T = 0 & \text{(zero twisting moment)} \\
& M = 0 & \text{(zero bending moment)} \\
& M' = 0 & \text{(zero shear force)}
\end{aligned} \tag{4.117}$$

For composite beams, the offsets d and e may be defined in a manner similar to the way they were defined for isotropic beams: d is the distance from the $\overline{y}$ axis to the cross-sectional mass centroid, positive when the mass centroid is toward the leading edge from the $\overline{y}$ axis; and e is the distance from the $\overline{y}$ axis to the aerodynamic center, positive when the aerodynamic center is toward the leading edge from the $\overline{y}$ axis. Recall that for composite beams, the $\overline{y}$ axis must have different properties from those it has for isotropic beams, and the term "elastic axis" has a different meaning. For a spanwise uniform isotropic beam, the elastic axis is along the $\overline{y}$ axis and is the locus of cross-sectional shear centers; transverse forces acting through this axis do not twist the beam. For spanwise uniform composite beams with bending-twist coupling, no axis can be defined as the locus of a cross-sectional property through which transverse shear forces can act without twisting the beam. For such beams, we must place the $\overline{y}$ axis along the locus of shear centers, a point in the cross section at which transverse shear forces are structurally decoupled from the twisting moment. Although transverse shear forces acting at the $\overline{y}$ axis do not *directly* induce twist, the bending moment induced by the shear force still induces twist when $K \neq 0$.

We can now write the homogeneous part of the equations of equilibrium as

$$\overline{\theta}'' - \frac{K}{GJ}w''' + \frac{qeca}{GJ}\theta\cos(\Lambda) = 0$$

$$w'''' - \frac{K}{EI}\overline{\theta}''' - \frac{qca}{EI}\theta\cos(\Lambda) = 0$$

(4.118)

Differentiating the first equation with respect to $\overline{y}$ and transforming the set of equations so that they are uncoupled in the highest derivative terms $\overline{\theta}'''$ and w'''', we obtain

$$\overline{\theta}''' + \frac{\overline{EI}\,\overline{GJ}}{\overline{EI}\,\overline{GJ} - K^2}\frac{qeca}{GJ}\theta'\cos(\Lambda) - \frac{K\,\overline{EI}}{\overline{EI}\,\overline{GJ} - K^2}\frac{qca}{EI}\theta\cos(\Lambda) = 0$$

$$w'''' + \frac{K\,\overline{GJ}}{\overline{EI}\,\overline{GJ} - K^2}\frac{qeca}{GJ}\theta'\cos(\Lambda) - \frac{\overline{EI}\,\overline{GJ}}{\overline{EI}\,\overline{GJ} - K^2}\frac{qca}{EI}\theta\cos(\Lambda) = 0$$

(4.119)

Multiplying the first equation by $\cos(\Lambda)$ and the second equation by $\sin(\Lambda)$ and subtracting the second equation from the first, we obtain a single equation in terms of $\theta = \overline{\theta}\cos(\Lambda) - w'\sin(\Lambda)$ as

$$\theta''' + \frac{\overline{EI}\,\overline{GJ}}{\overline{EI}\,\overline{GJ} - K^2}\frac{qeca\ell^2}{GJ}\cos^2(\Lambda)\left[1 - \frac{K}{EI}\tan(\Lambda)\right]\theta'$$

$$+ \frac{\overline{EI}\,\overline{GJ}}{\overline{EI}\,\overline{GJ} - K^2}\frac{qca\ell^3}{EI}\sin(\Lambda)\cos(\Lambda)\left[1 - \frac{K}{GJ}\frac{1}{\tan(\Lambda)}\right]\theta = 0$$

(4.120)

where $(\ )'$ now denotes $d(\ )/d\eta$ as in the parallel development for the elastically uncoupled wing discribed previously.

The boundary conditions can be derived from Eqs. (4.117) as

$$\theta(0) = \theta'(1) = \theta''(1) + \frac{\overline{EI}\,\overline{GJ}}{\overline{EI}\,\overline{GJ} - K^2}\frac{qeca\ell^2}{GJ}\cos^2(\Lambda)\left[1 - \frac{K}{EI}\tan(\Lambda)\right]\theta(1) = 0$$

(4.121)

The aeroelastic divergence problem with structural coupling has the same mathematical form as the problem without coupling, an approximate solution of which is given in the previous section. We can see that the parameters τ and β can be redefined as

$$\tau = \frac{\overline{EI}\,\overline{GJ}}{\overline{EI}\,\overline{GJ} - K^2}\frac{qeca\ell^2}{GJ}\cos^2(\Lambda)\left[1 - \frac{K}{EI}\tan(\Lambda)\right]$$

$$\beta = \frac{\overline{EI}\,\overline{GJ}}{\overline{EI}\,\overline{GJ} - K^2}\frac{qca\ell^3}{EI}\sin(\Lambda)\cos(\Lambda)\left[1 - \frac{K}{GJ}\frac{1}{\tan(\Lambda)}\right]$$

(4.122)

and, again, the divergence boundary can be expressed approximately in terms of the line

$$\tau_D = \frac{\pi^2}{4} + \frac{3\pi^2}{76}\beta_D$$

(4.123)

Using the expressions for the parameters in the equation of the divergence boundary, we have

$$q_D = \frac{\pi^2}{4} \frac{\overline{EI}\,\overline{GJ} - K^2}{\overline{EI}\,\overline{GJ}} \frac{\overline{GJ}}{eca\ell^2 \cos^2(\Lambda)} \frac{1}{\left\{1 - \frac{K}{\overline{EI}}\tan(\Lambda) - \frac{3\pi^2}{76}\frac{\ell}{e}\frac{\overline{GJ}}{\overline{EI}}\left[\tan(\Lambda) - \frac{K}{\overline{GJ}}\right]\right\}} \tag{4.124}$$

We can simplify this by introducing the dimensionless parameter

$$\kappa = \frac{K}{\sqrt{\overline{EI}\,\overline{GJ}}} \tag{4.125}$$

so that

$$q_D = \frac{\pi^2\overline{GJ}(1 - \kappa^2)}{4eca\ell^2 \cos^2(\Lambda)\left\{1 - \kappa\sqrt{\frac{\overline{GJ}}{\overline{EI}}}\tan(\Lambda) - \frac{3\pi^2}{76}\frac{\ell}{e}\frac{\overline{GJ}}{\overline{EI}}\left[\tan(\Lambda) - \kappa\sqrt{\frac{\overline{EI}}{\overline{GJ}}}\right]\right\}} \tag{4.126}$$

With Eq. (4.126), we may determine the divergence dynamic pressure with sufficient accuracy to ascertain its trends versus sweep angle Λ and elastic-coupling parameter κ. The formula shows that there is a strong relationship between these two quantities.

To illustrate the utility of this analysis, let us first normalize q_D with the value it would have at zero sweep angle and zero coupling—namely, q_{D_0}, so that

$$\frac{q_D}{q_{D_0}} = \frac{(1 - \kappa^2)\left[1 + \tan^2(\Lambda)\right]}{1 - \kappa\sqrt{\frac{\overline{GJ}}{\overline{EI}}}\tan(\Lambda) - \frac{3\pi^2}{76}\frac{\ell}{e}\frac{\overline{GJ}}{\overline{EI}}\left[\tan(\Lambda) - \kappa\sqrt{\frac{\overline{EI}}{\overline{GJ}}}\right]} \tag{4.127}$$

As before, when the denominator of the expression for divergence dynamic pressure vanishes, it corresponds to infinite divergence dynamic pressure; crossing this "boundary" means crossing from a regime in which divergence occurs to one in which it does not. Setting the denominator to zero and solving for the tangent of the sweep angle, we obtain

$$\tan(\Lambda_\infty) = \frac{1 + \frac{3\pi^2}{76}\sqrt{\frac{\overline{GJ}}{\overline{EI}}}\frac{\ell}{e}\kappa}{\frac{3\pi^2}{76}\frac{\overline{GJ}}{\overline{EI}}\frac{\ell}{e} + \sqrt{\frac{\overline{GJ}}{\overline{EI}}}\kappa} \tag{4.128}$$

where Λ_∞ is the sweep angle at which the divergence dynamic pressure goes to infinity. With this definition, we can rewrite Eq. (4.127) as

$$\frac{q_D}{q_{D_0}} = \frac{(1 - \kappa^2)\left[1 + \tan^2(\Lambda)\right]}{\left(1 + \frac{3\pi^2}{76}\sqrt{\frac{\overline{GJ}}{\overline{EI}}}\frac{\ell}{e}\kappa\right)\left[1 - \frac{\tan(\Lambda)}{\tan(\Lambda_\infty)}\right]} \tag{4.129}$$

Again, divergence is possible only if $-90° < \Lambda < \Lambda_\infty$. Thus, because of the presence of κ as an additional design parameter, designers can at least partially compensate for the destabilizing effect of forward sweep by appropriately choosing $\kappa < 0$, which for an increment of upward bending of the wing provides an increment of nose-down twisting. There is a limit to how much coupling can be achieved, however, because typically $|\kappa| < 0.86$.

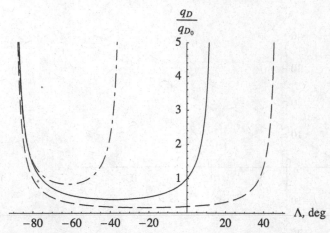

Figure 4.28. Normalized divergence dynamic pressure for an elastically coupled, swept wing with $\overline{GJ/EI} = 0.2$ and $e/\ell = 0.02$; $\kappa = -0.4$ (dots and dashes), $\kappa = 0$ (solid lines), $\kappa = 0.4$ (dashed lines)

There are two main differences between the designs of isotropic and composite wings. First, it is possible to achieve a much wider range of values for $\overline{GJ/EI}$. Second—and significantly more powerful—is the fact that composite wings can be designed with nonzero values of κ. From Eq. (4.128), the value of Λ_∞ is decreased as κ is decreased, which means that the range of Λ over which divergence occurs is decreased. To confirm this and the previous statement about positive κ being destabilizing, Fig. 4.28 shows results for $\kappa = -0.4$, 0, and 0.4. It is clear that a composite wing can be swept forward and still avoid divergence with a proper choice (i.e., a sufficiently large and negative value) of κ. Because forward sweep has advantages for the design of highly maneuverable aircraft, this is a result of practical importance. The sweep angles at which divergence becomes impossible, Λ_∞, are also somewhat sensitive to $\overline{GJ/EI}$ and e/ℓ, as shown in Figs. 4.29 and 4.30. Evidently, divergence-free, forward-swept wings may be designed with larger sweep angles by decreasing torsional stiffness relative to bending stiffness and by decreasing e/ℓ.

4.3 Epilogue

In this chapter, we considered divergence and aileron reversal of simple wind-tunnel models; torsional divergence, load redistribution, and aileron reversal in flexible-beam representations of lifting surfaces; roll effectiveness of an airplane with wings modeled as beams; the effects of sweep on coupled bending-torsion divergence; and the role of aeroelastic tailoring. It is clear that aircraft design is strongly influenced by aeroelastic considerations. In all of the cases explored in this chapter, the inertial loads are inconsequential and therefore were neglected. In Chapter 5, inertial loads are introduced into the aeroelastic analysis of flight vehicles, and the flutter problem is explored.

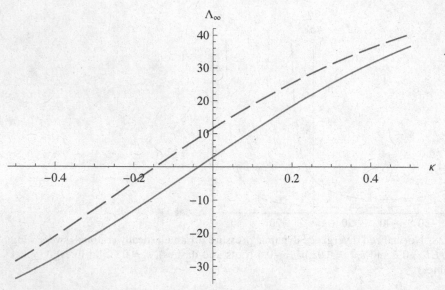

Figure 4.29. Sweep angle for which divergence dynamic pressure is infinite for a wing with $\overline{GJ/EI} = 0.5$; solid line is for $e/\ell = 0.01$; dashed line is for $e/\ell = 0.04$

Problems

1. Consider a rigid, wind-tunnel model of a uniform wing, which is pivoted in pitch about the mid-chord and elastically restrained in pitch by a linear spring with spring constant of 225 lb/in mounted at the trailing edge. The model has a symmetric airfoil, a span of 3 feet, and a chord of 6 inches. The total lift-curve

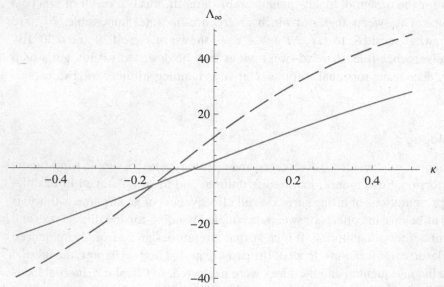

Figure 4.30. Sweep angle for which divergence dynamic pressure is infinite for a wing with $e/\ell = 0.02$; solid line is for $\overline{GJ/EI} = 1.0$; dashed line is for $\overline{GJ/EI} = 0.25$

slope is 6 per rad. The aerodynamic center is located at the quarter-chord, and the mass centroid is at the mid-chord.

(a) Calculate the divergence dynamic pressure at sea level.

(b) Calculate the divergence airspeed at sea level.

Answers: $q_D = 150 \text{ lb/ft}^2$; $U_D = 355$ ft/sec

2. For the model in Problem 1, for a dynamic pressure of 30 lb/ft^2, compute the percentage change in lift caused by the aeroelastic effect.

Answer: 25%

3. For the model of Problem 1, propose design changes in the support system that would double the divergence dynamic pressure by

(a) changing the stiffness of the restraining spring

(b) relocating the pivot point

Answers: (a) $k = 450$ lb/in; (b) $x_O = 2.513$ in

4. For the model of Problem 1 as altered by the design changes of Problem 3, calculate the percentage change in lift caused by the aeroelastic effect for a dynamic pressure of 30 lb/ft^2, a weight of 3 lb, $\alpha_r = 0.5°$, and for

(a) the design change of Problem 3a

(b) the design change of Problem 3b

Answers: 11.11%; 17.91%

5. Consider a strut-mounted wing similar to the one discussed in Section 4.1.3, except that the two springs may have different stiffnesses. Denoting the leading-edge spring constant by k_1 and the trailing-edge spring constant by k_2, and assuming that the aerodynamic center is at the quarter-chord, show that divergence can be eliminated if $k_1/k_2 \geq 3$.

6. Using Excel or a similar tool, plot a family of curves that depict the relationship of the aileron-elastic efficiency, η, versus normalized dynamic pressure, $\bar{q} = q/q_D$, for various values of $R = q_R/q_D$ and $0 < \bar{q} < 1$. Make two plots on the following scales to reduce confusion:

(a) Plot $R < 1$ using axes $-3 < \eta < 3$

(b) Plot $R > 1$ using axes $-3 < \eta < 3$

Hint: Do not compute values for the cases in which $1 < R < 1.1$; Excel does not handle these well and you may get confused. For some cases, you may want to plot symbols only and nicely sketch the lines that form the curves.

Answer the following questions: Where does aileron reversal occur? If you had to design a wing, what R would you try to match (or approach) and why? What happens when $q_R = q_D$? How does the efficiency change as q approaches q_R? Why do you think this happens? What other pertinent features can you extract from these plots? Explain how you came to these conclusions.

7. Consider a torsionally elastic ($\overline{GJ} = 8,000$ lb in^2) wind-tunnel model of a uniform wing, the ends of which are rigidly fastened to the wind-tunnel walls. The model has a symmetric airfoil, a span of 3 feet, and a chord of 6 inches.

The sectional lift-curve slope is 6 per rad. The aerodynamic center is located at the quarter-chord, and both the mass centroid and the elastic axis are at the mid-chord.

(a) Calculate the divergence dynamic pressure at sea level.

(b) Calculate the divergence airspeed at sea level.

Answers: (a) $q_D = 162.46$ lb/ft^2; (b) $U_D = 369.65$ ft/sec

8. For the model in Problem 7, propose design changes in the model that would double the divergence dynamic pressure by

(a) changing the torsional stiffness of the wing

(b) relocating the elastic axis

Answers: $\overline{GJ} = 16{,}000$ lb in^2; $x_{ea} = 2.25$ in

9. For the model of Problem 7, for a dynamic pressure of 30 lb/ft^2, compute the percentage increase in the sectional lift at mid-span caused by the aeroelastic effect.

Answer: 28.09%

10. For the model in Problem 7, for a dynamic pressure of 30 lb/ft^2, compute the percentage increase in the total lift caused by the aeroelastic effect.

Answer: 18.58%

11. Consider a swept clamped-free wing, as described in Section 4.2.6. The governing partial differential equations are given in Eqs. (4.94) and the boundary conditions in Eqs. (4.95). An approximate solution is sought for a wing with a symmetric airfoil, using a truncated set of assumed modes and the generalized equations of equilibrium: a specialized version of the generalized equations of motion for which all time-dependent terms are zero. Note that what is being asked for here is equivalent to the application of the Ritz method to the principle of virtual work (see Section 3.5). With the wing weight ignored, only structural and aerodynamic terms are involved. The structural terms of the generalized equations of equilibrium are based on the potential energy (here, the strain energy) given by

$$P = \frac{1}{2} \int_0^\ell \left(\overline{EI} w''^2 - 2K w'' \overline{\theta}' + \overline{GJ} \,\overline{\theta}'^2 \right) d\bar{y}$$

and the bending and torsion deformation is represented in terms of a truncated series, such that

$$w = \sum_{i=1}^{N_w} \eta_i \Psi_i(\bar{y})$$

$$\overline{\theta} = \sum_{i=1}^{N_\theta} \phi_i \Theta_i(\bar{y})$$

where N_w and N_θ are the numbers of assumed modes used to represent bending and torsion, respectively; η_i and ϕ_i are the generalized coordinates associated with bending and torsion, respectively; and Ψ_i and Θ_i are the assumed mode shapes for bending and torsion, respectively. Determine the potential energy in terms of the generalized coordinates using as assumed modes the uncoupled, clamped-free, free-vibration modes of torsion and bending. For torsion

$$\Theta_i = \sqrt{2}\sin\left[\frac{\pi\left(i - \frac{1}{2}\right)\bar{y}}{\ell}\right]$$

For bending, according to Eq. (3.258), Ψ_i is given as

$$\Psi_i = \cosh(\alpha_i\bar{y}) - \cos(\alpha_i\bar{y}) - \beta_i[\sinh(\alpha_i\bar{y}) - \sin(\alpha_i\bar{y})]$$

with α_i and β_i as given in Table 3.1.

12. Work Problem 11, but for assumed modes, instead of using the expressions given therein, use

$$\Theta_i = \left(\frac{\bar{y}}{\ell}\right)^i$$

$$\Psi_i = \left(\frac{\bar{y}}{\ell}\right)^{i+1}$$

recalling that these functions are not orthogonal.

13. Work Problem 11, but use the finite element method to represent both bending and torsion.

14. Referring to Problem 11, 12, or 13, starting with the virtual work of the aerodynamic forces as

$$\overline{\delta W} = \int_0^\ell \left(L'\delta w + M'\delta\theta\right)d\bar{y}$$

where L' and M' are the sectional lift and pitching-moment expressions used to develop Eqs. (4.94), assuming a symmetric airfoil and using the given deformation modes, find the generalized forces Ξ_i, $i = 1, 2, \ldots, N = N_w + N_\theta$. As discussed in the text, generalized forces are the coefficients of the variations of the generalized coordinates in the virtual-work expression. (Hint: Neglecting the weight terms on the right-hand sides of Eqs. 4.92, we find L' is the right-hand side of the second of those equations, whereas M' is the negative of the right-hand side of the first and equal to eL'.)

15. Referring to Problems 14 and 11, 12, or 13, determine the generalized equations of equilibrium in the form

$$[K]\{\xi\} = \bar{q}\left\{[A]\{\xi\} + \{\Xi_0\}\right\}$$

where $\bar{q}$ is the dimensionless dynamic pressure given by q/q_{D_0}; q_{D_0} is the torsional divergence dynamic pressure of the unswept clamped-free wing, given by Eq. (4.55); $\{\xi\}$ is the column matrix of all unknowns $\bar{\eta}_i = \eta_i/\ell$, $i = 1, 2, \ldots, N_w$, and ϕ_i, $i = 1, 2, \ldots, N_\theta$; and $\{\Xi_0\}$ is an $N \times 1$ column matrix containing the parts of the aerodynamic generalized forces that do not depend on any unknowns; and $N = N_w + N_\theta$. (Note that in application of the finite element, $N_w = N_\theta$ refers to the number of finite elements, and $N = 2N_w + N_\theta$.) The $N \times N$ matrices $[K]$

and $[A]$ are the stiffness and aerodynamic matrices, respectively. If Problem 11 is the basis for solution and elastic coupling is ignored, then the stiffness matrix $[K]$ is diagonal because the normal modes used to represent the wing structural behavior are orthogonal with respect to the stiffness properties of the wing.

16. Referring to Problem 15, perform the following numerical studies:

   a. *Divergence*: To determine the divergence dynamic pressure, write the *homogeneous* generalized equations of equilibrium in the form

$$\frac{1}{q}\{\xi\} = [K]^{-1}[A]\{\xi\}$$

   which is obviously an eigenvalue problem with $1/\overline{q}$ as the eigenvalue. After solving the eigenvalue problem, the largest $1/\overline{q}$ provides the lowest dimensionless critical divergence dynamic pressure $\overline{q}_D = q_D/q_{D_0}$ at the sweep angle under consideration. By numerical experimentation, determine N_w and N_θ to obtain the divergence dynamic pressure to within plotting accuracy. Plot the divergence dynamic pressure versus sweep angle for a range of values for the sweep angle $-45° \leq \Lambda \leq 45°$ and values of the dimensionless parameters e/ℓ (0.05 and 0.1), $\overline{EI/GJ}$ (1 and 5), and $\kappa = 0, \pm 0.5$. Compare your results with those obtained in Eq. (4.129). Comment on the accuracy of the approximate solution in the text versus your Ritz or finite element solution. Which one should be more accurate? Discuss the trends of divergence dynamic pressure that you see regarding the sweep angle, stiffness ratio, and location of the aerodynamic center.

   b. *Response*: For the response, you need to consider the inhomogeneous equations, which should be put into the form

$$[[K] - \overline{q}[A]]\{\xi\} = \overline{q}\{\Xi_0\}$$

   Letting $\alpha_r = 1°$, obtain the response by solving the linear system of equations represented in this matrix equation. Plot the response of the wing tip (i.e., w and $\overline{\theta}$ at $\overline{y} = \ell$) for varying dynamic pressures up to $q = 0.95q_D$ for the above values of e/ℓ, $\overline{EI/GJ}$, and κ with $\Lambda = -25°$ and $0°$. Plot the lift, twist, and bending-moment distributions for the case with the largest tip-twist angle. Comment on this result and on the trends of static-aeroelastic response that you see regarding the sweep angle, stiffness ratio, location of the aerodynamic center, and elastic coupling.

17. Consider the divergence of an unswept composite wing with $\kappa = 0$, $\overline{GJ/EI} = 0.2$, and $e/\ell = 0.025$. Using Eq. (4.126), determine the value of κ, as defined by Eq. (4.125), needed to keep the divergence dynamic pressure unchanged for forward-swept wings with various values of $\Lambda < 0$. Plot these values of κ versus Λ.

18. Using the approximate formula found in Eq. (4.126), derive a formula for $q_D ac\ell^3/\overline{EI}$ analogous to Eq. (4.113) and use it to determine the divergence dynamic pressure for swept composite wings when $e < 0$. Discuss the situations in which we might encounter a negative value of e. Which sign of κ would you

expect to be stabilizing in this case? Plot this normalized divergence dynamic pressure for a swept composite wing with $\overline{GJ}/\overline{EI} = 0.2$ and $e/\ell = -0.025$ versus Λ for $\kappa = 0$ and ± 0.4.

19. Consider the divergence of a swept composite wing. Show that the governing equation and boundary conditions found in Eqs. (4.120) and (4.121) can be written as a second-order, integro-differential equation of the form

$$\theta'' + \tau\theta - r\tau \int_\eta^1 \theta(\xi)d\xi = 0$$

with boundary conditions $\theta(0) = \theta'(1) = 0$ and with $r = \beta/\tau$. Determine the two simplest polynomial comparison functions for this reduced-order equation and boundary conditions. Use Galerkin's method to obtain one- and two-term approximations to the divergence dynamic pressure τ_D versus r. Plot your approximate solutions for the case in which $\overline{GJ}/\overline{EI} = 0.2$, $e/\ell = 0.02$, and $\kappa = -0.4$, depicted in Fig. 4.28, and compare these with the approximate solution given in the text. For the two-term approximation, determine the limit point for positive e, noting that the exact values are $r = 1.59768$ and $\tau_D = 10.7090$.

Answer: The one-term approximation is

$$\tau_D = \frac{30}{12 - 5r}$$

The two-term approximation is

$$\tau_D = \frac{1260}{282 - 105r \pm \sqrt{3}\sqrt{15r(197r - 1{,}036) + 17{,}408}}$$

The approximate limit point in the first quadrant is at $r = 1.61804$ and $\tau_D = 11.2394$. Within plotting accuracy, the two-term approximation is virtually indistinguishable from the exact solution when $-10 \le \tau_D \le 10$.

20. Consider a uniform, torsionally flexible wing of length ℓ with torsional stiffness $\overline{GJ}$ and with the aileron extending from $y = r\ell$ to $y = R\ell$.

   (a) Find the expression that must be solved for $\lambda\ell$ using the criterion for aileron reversal that the change in root-bending moment with respect to aileron deflection vanish. Use a tip-loss factor of B.

   (b) Assuming that $e = 0.25c$, $c_{\ell_\beta} = 0.8$, $c_{m_\beta} = -0.5$, and $R = 1$, and considering both $r = 0$ and $r = 0.5$ and $B = 0.97$ and $B = 1$, find $\lambda_1\ell$. Discuss the effect of r and B.

   (c) Determine $\lambda_1\ell$ from Eq. (4.79). Comparing this to your results for the case of $r = 0$, $R = 1$, and $B = 1$, explain how it is possible for the reversal dynamic pressure extracted from the bending-moment criterion to be different from that extracted for the total lift criterion.

Answer: For example, when $r = 0$, $R = 1$, and $B = 1$: $\sec(\lambda\ell) = \dfrac{e\,c_{\ell_\beta} + c\,c_{m_\beta}\left[1 + \frac{(\lambda\ell)^2}{2}\right]}{e\,c_{\ell_\beta} + c\,c_{m_\beta}}$; $\lambda_1\ell = 0.984774$

21. Consider a rigid body that represents the fuselage of a symmetric aircraft, to which are attached uniform, torsionally flexible wings that have the same properties as those in Problem 20. Assuming the aircraft is flying at constant speed with a constant roll rate, develop solutions for the same sets of parameters as asked for in Problem 20.

5 Aeroelastic Flutter

> The pilot of the airplane...succeeded in landing with roughly two-thirds of his horizontal tail surface out of action; some others have, unfortunately, not been so lucky.... The flutter problem is now generally accepted as a problem of primary concern in the design of current aircraft structures. Stiffness criteria based on flutter requirements are, in many instances, the critical design criteria.... There is no evidence that flutter will have any less influence on the design of aerodynamically controlled booster vehicles and re-entry gliders than it has, for instance, on manned bombers.
>   —R. L. Bisplinghoff and H. Ashley in *Principles of Aeroelasticity*, John Wiley and Sons, Inc., 1962

Chapter 3 addressed the subject of structural dynamics, which is the study of phenomena associated with the interaction of inertial and elastic forces in mechanical systems. In particular, the mechanical systems considered were one-dimensional, continuous configurations that exhibit the general structural-dynamic behavior of flight vehicles. If in the analysis of these structural-dynamic systems aerodynamic loading is included, then the resulting dynamic phenomena may be classified as aeroelastic. As observed in Chapter 4, aeroelastic phenomena can have a significant influence on the design of flight vehicles. Indeed, these effects can greatly alter the design requirements that are specified for the disciplines of performance, structural loads, flight stability and control, and even propulsion. In addition, aeroelastic phenomena can introduce catastrophic instabilities of the structure that are unique to aeroelastic interactions and can limit the flight envelope.

Recalling the diagram in Fig. 1.1, we can classify aeroelastic phenomena as either static or dynamic. Whereas Chapter 4 addressed only static aeroelasticity, in this chapter, we examine dynamic aeroelasticity. Although there are many other dynamic aeroelastic phenomena that could be treated, we focus entirely on the instability called "flutter," which generally leads to a catastrophic structural failure of a flight vehicle. A formal definition of aeroelastic flutter is as follows: *a dynamic instability of a flight vehicle associated with the interaction of aerodynamic, elastic, and inertial forces.* From this definition, it is apparent that any investigation of flutter stability requires an adequate knowledge of the system's structural dynamic and aerodynamic properties. To further elaborate, flutter is a self-excited and potentially destructive oscillatory instability in which aerodynamic forces on a flexible body couple with its natural modes of vibration to produce oscillatory motions with increasing amplitude. In such cases, the level of vibration will increase, resulting in oscillatory motion with amplitude sufficiently large to cause structural failure.

Because of this, structures exposed to aerodynamic forces—including wings and airfoils but also chimneys and bridges—must be carefully designed to avoid flutter. In complex systems in which neither the aerodynamics nor the mechanical properties are fully understood, the elimination of flutter can be guaranteed only by thorough testing. Of the various phenomena that are categorized as aeroelastic flutter, lifting-surface flutter is most often encountered and most likely to result in a catastrophic structural failure. As a result, it is required that lifting surfaces of all flight vehicles be analyzed and tested to ensure that this dynamic instability will not occur for any condition within the vehicle's flight envelope.

If the airflow about the lifting surface becomes separated during any portion of an unstable oscillatory cycle of the angle of attack, the governing equations become nonlinear and the instability is referred to as "stall flutter." Stall flutter most commonly occurs on turbojet compressor and helicopter rotor blades. Other phenomena that result in nonlinear behavior include large deflections, mechanical slop, and nonlinear control systems. Nonlinear phenomena are not considered in the present treatment. Even with this obvious paring down of the problem, however, we still find that linear-flutter analysis of clean lifting surfaces is complicated. Thus, we can offer only a simplified discussion of the theory of flutter. Readers are urged to consult the references for additional information on the subject.

This chapter begins by using the modal representation to set up a lifting-surface flutter analysis as a linear set of ordinary differential equations. These are transformed into an eigenvalue problem, and the stability characteristics are discussed in terms of the eigenvalues. Then, as an example of this methodology, a two-degree-of-freedom "typical-section" analysis is formulated using the simple steady-flow aerodynamic model used in Chapter 4. The main shortcoming of this simple analysis is the neglect of unsteady effects in the aerodynamic model. Motivated by the need to consider unsteady aerodynamics in a meaningful but simple way, we then introduce classical flutter analysis. Engineering solutions that partially overcome the shortcomings of classical flutter analysis follow. To complete the set of analytical tools needed for flutter analysis, two different unsteady-aerodynamic theories are outlined: one suitable for use with classical flutter analysis and its derivatives; the other suitable for eigenvalue-based flutter analysis. After illustrating how to approach the flutter analysis of a flexible wing using the assumed-modes method, the chapter concludes with a discussion of flutter-boundary characteristics.

5.1 Stability Characteristics from Eigenvalue Analysis

The lifting-surface flutter of immediate concern can be described by a linear set of structural dynamic equations that include a linear representation of the unsteady airloads in terms of the elastic deformations. The surface could correspond to a wing or stabilizer either with or without control surfaces. Analytical simulation of the surface is sometimes made more difficult by the presence of external stores, engine nacelles, landing gear, or internal fuel tanks. Although such complexities complicate the analysis, they do not alter significantly the physical character of the flutter instability. For this reason, the following discussion is limited to a "clean" lifting surface.

When idealized for linear analysis, the nature of flutter is such that the flow over the lifting surface creates not only steady components of lift and pitching moment but also dynamic forces in response to small perturbations of the lifting-surface motion. The wing airfoil at a local cross section undergoes pitch and plunge motions from lifting-surface torsional and bending deformation, respectively. When a lifting surface that is statically stable below its flutter speed is disturbed, the oscillatory motions caused by those disturbances die out in time with exponentially decreasing amplitudes. That is, we could say that the air provides damping for all such motions. Above the flutter speed, however, rather than damping out the motions due to small perturbations in the configuration, the air can be said to provide negative damping. Thus, these oscillatory motions grow with exponentially increasing amplitudes. This qualitative description of flutter can be observed in a general discussion of stability characteristics based on complex eigenvalues.

Before attempting to conduct an analysis of flutter, it is instructive to first examine the possible solutions to a structural-dynamic representation in the presence of airloads. We presume that a flight vehicle can be represented in terms of its normal modes of vibration. We illustrate this with the lifting surface modeled as a plate rather than a beam. This is more realistic for low-aspect-ratio wings but, in the present framework, this increased realism presents little increase in complexity because of the modal representation. For displacements $w(x, y, t)$ in the z direction normal to the plane of the planform (i.e., the x-y plane), the normal mode shapes can be represented by $\phi_i(x, y)$ and the associated natural frequencies by ω_i. A typical displacement of the structure can be written as

$$w(x, y, t) = \sum_{i=1}^{n} \xi_i(t)\phi_i(x, y) \tag{5.1}$$

where $\xi_i(t)$ is the generalized coordinate of the ith mode. For simplicity, both rigid-body and elastic modes are included in this set without special notation to distinguish them from one another. The set of generalized equations of motion for the flight vehicle can be written as

$$M_i(\ddot{\xi}_i + \omega_i^2 \xi_i) = \Xi_i \qquad (i = 1, 2, \ldots, n) \tag{5.2}$$

where M_i is the generalized mass associated with the mass distribution $m(x, y)$ and can be determined as

$$M_i = \iint_{\text{planform}} m(x, y)\phi_i^2(x, y)dxdy \tag{5.3}$$

The generalized force $\Xi_i(t)$, associated with the external loading $F(x, y, t)$, can be evaluated as

$$\Xi_i(t) = \iint\limits_{\text{planform}} F(x, y, t)\phi_i(x, y)dxdy \qquad (5.4)$$

Recall that of the set of natural frequencies ω_i, any that are associated with rigid-body modes are equal to zero.

To examine the stability properties of the flight vehicle, the only external loading to be considered is from the aerodynamic forces, which can be represented as a linear function of $w(x, y, t)$ and its partial derivatives, plus a set of additional states that may be needed to represent pertinent aspects of the flow field, such as the induced flow or downwash. It is presumed that all other external disturbances have been eliminated. Such external disturbances normally would include atmospheric gusts, store-ejection reactions, and so forth. Recalling that the displacement can be represented as a summation of the modal contributions, the induced-pressure distribution, $\Delta p(x, y, t)$, can be described as a linear function of the generalized coordinates, their derivatives, and the flow-field states. Such a relationship can be written as

$$\Delta p(x, y, t) = \sum_{j=1}^{n} \left[a_j(x, y)\xi_j(t) + b_j(x, y)\dot{\xi}_j(t) + c_j(x, y)\ddot{\xi}_j(t) \right] + \sum_{j=1}^{N} d_j(x, y)\lambda_j(t)$$

$$(5.5)$$

where the λs are state variables associated with the flow field, sometimes called "augmented" states or "lag" states, written here so as to have the same units as the generalized coordinates. The number of these states is denoted by $N \geq 0$, which may be distinct from n. The corresponding generalized force of the ith mode now can be determined from

$$\Xi_i(t) = \iint\limits_{\text{planform}} \Delta p(x, y, t)\phi_i(x, y)dxdy$$

$$= \sum_{j=1}^{n} \xi_j(t) \iint\limits_{\text{planform}} a_j(x, y)\phi_i(x, y)dxdy$$

$$+ \sum_{j=1}^{n} \dot{\xi}_j(t) \iint\limits_{\text{planform}} b_j(x, y)\phi_i(x, y)dxdy$$

$$+ \sum_{j=1}^{n} \ddot{\xi}_j(t) \iint\limits_{\text{planform}} c_j(x, y)\phi_i(x, y)dxdy \qquad (5.6)$$

$$+ \sum_{j=1}^{N} \lambda_j(t) \iint\limits_{\text{planform}} d_j(x, y)\phi_i(x, y)dxdy$$

$$= \rho_\infty \frac{U^2}{b^2} \left[\sum_{j=1}^{n} \left(a_{ij}\xi_j + \frac{b}{U}b_{ij}\dot{\xi}_j + \frac{b^2}{U^2}c_{ij}\ddot{\xi}_j \right) + \sum_{j=1}^{N} d_{ij}\lambda_j \right]$$

Following the convention in some published work, we factored out the freestream air density ρ_∞ and U^2/b^2 from the aerodynamic generalized force expression. Although not necessary, this step enables analysts to identify altitude effects more readily. It also shows explicitly that all aerodynamic effects vanish in a vacuum where ρ_∞ vanishes. Moreover, the normalization involving powers of b/U—where b is a reference semi-chord of the lifting surface—allows the matrices $[a]$, $[b]$, $[c]$, and $[d]$ to have the same units, simplifying the equations in terms of dimensionless variables that follow later. Any inhomogeneous terms in the generalized forces can be eliminated by redefinition of the generalized coordinates so that they are measured with respect to a different reference configuration. Thus, the generalized equations of motion can be written as a homogeneous set of differential equations when this form of the generalized force is included. They are

$$\frac{b^2}{U^2}\left(M_i\ddot{\xi}_i + M_i\omega_i^2\xi_i\right) - \rho_\infty\frac{b^2}{U^2}\sum_{j=1}^{n}c_{ij}\ddot{\xi}_j - \rho_\infty\frac{b}{U}\sum_{j=1}^{n}b_{ij}\dot{\xi}_j$$

$$-\rho_\infty\sum_{j=1}^{n}a_{ij}\xi_j - \rho_\infty\sum_{j=1}^{N}d_{ij}\lambda_j = 0 \quad (i = 1, 2, \ldots, n) \tag{5.7}$$

If $N > 0$, then N additional equations are needed for the λs. Such equations generally have the form

$$\sum_{j=1}^{N}A_{ij}\dot{\lambda}_j + \frac{U}{b}\left(\lambda_i - \sum_{j=1}^{n}E_{ij}\xi_j\right) = 0 \quad (i = 1, 2, \ldots, N) \tag{5.8}$$

or

$$[A]\{\dot{\lambda}\} + \frac{U}{b}\left\{\{\lambda\} - [E]\{\xi\}\right\} = 0 \tag{5.9}$$

Matrices $[A]$ and $[E]$ can be obtained from unsteady-aerodynamic theories as well as from computational fluid dynamics or test data. Note that matrix $[E]$ may be an operator that differentiates $\{\xi\}$ one or more times.

This system consists of $n + N$ equations—that is, the number of structural modes (including both elastic and rigid-body modes) plus the number of aerodynamic states, respectively. The general solution to this set of linear ordinary differential equations can be described as a simple exponential function of time because they are homogeneous. The form of this solution is taken as

$$\xi_i(t) = \overline{\xi}_i \exp(\nu t) \qquad \lambda_i(t) = \overline{\lambda}_i \exp(\nu t) \tag{5.10}$$

Substituting this expression into Eqs. (5.7) and (5.8), we obtain a set of algebraic equations, each term of which contains $\exp(\nu t)$. After factoring out this term, the result is $n + N$ simultaneous linear, homogeneous, algebraic equations for the $\overline{\xi}$s, which may be written in matrix form as

$$\left[p^2\lceil M.\rceil + \frac{b^2}{U^2}\lceil M\omega^2.\rceil\right]\{\overline{\xi}\} - \rho_\infty\left[p^2[c] + p[b] + [a]\right]\{\overline{\xi}\} - \rho_\infty[d]\{\overline{\lambda}\} = \{0\}$$

$$[p[A] + [I]]\{\overline{\lambda}\} - [E]\{\overline{\xi}\} = \{0\} \tag{5.11}$$

where $p = bv/U$ is the unknown dimensionless eigenvalue, and the symbol $\lceil M \cdot \rceil$ denotes a diagonal matrix with elements M_i. For a nontrivial solution of the generalized coordinate amplitudes, the determinant of the array formed by the coefficients of $\overline{\xi}_i$ and $\overline{\lambda}_i$ must be zero. It is apparent that this determinant is a polynomial of degree $2n + N$ in p. The subsequent solution of this polynomial equation for p yields $2n + N$ roots consisting of n_c complex conjugate pairs and n_r real numbers where $2n_c + n_r = 2n + N$. A typical complex root has the form

$$v_k = \frac{U p_k}{b} = \Gamma_k \pm i\Omega_k \qquad k = 1, 2, \ldots, n_c \tag{5.12}$$

whereas the roots v_k with $k = n_c + 1, n_c + 2, \ldots, n_c + n_r$ are real. In other words, any root can be written as v_k so that, $\Omega_k = 0$ for $n_c < k \leq n_c + n_r$.

For each root p_k, there are corresponding complex column matrices $\overline{\xi}_j^{(k)}$, $j = 1, 2, \ldots, n$, and $\overline{\lambda}_j^{(k)}$, $j = 1, 2, \ldots, N$. Thus, the solution for the displacement field from the generalized equations of motion with the aerodynamic coupling can be written as

$$w(x, y, t) = \sum_{k=1}^{n_c + n_r} \left\{ w_k(x, y) \exp\left[(\Gamma_k + i\Omega_k)t\right] + \overline{w}_k(x, y) \exp\left[(\Gamma_k - i\Omega_k)t\right] \right\} \tag{5.13}$$

where $\overline{w}_k$ is the complex conjugate of w_k. This expression for $w(x, y, t)$ turns out to be real, as expected. Each w_k represents a unique linear combination of the mode shapes of the structure; viz.

$$w_k(x, y) = \sum_{i=1}^{n} \overline{\xi}_i^{(k)} \phi_i(x, y) \qquad (k = 1, 2, \ldots, n_c + n_r) \tag{5.14}$$

Note that only the relative values of $\overline{\xi}_i^{(k)}$ can be determined unless the initial displacement and rate of displacement are specified.

It is apparent from the general solution for $w(x, y, t)$, Eq. (5.13), that the kth component of the summation represents a simple harmonic oscillation that is modified by an exponential function. The nature of this dynamic response to any specified initial condition is strongly dependent on the sign of each Γ_k. Typical response behavior is illustrated in Fig. 5.1 for positive, zero, and negative values of Γ_k when Ω_k is nonzero. We note that the negative of Γ_k is sometimes called the "modal damping" of the kth mode, and Ω_k is called the "modal frequency." It is also possible to classify these motions from the standpoint of stability. The convergent oscillations when $\Gamma_k < 0$ are termed "dynamically stable" and the divergent oscillations for $\Gamma_k > 0$ are "dynamically unstable." The case of $\Gamma_k = 0$ represents the boundary between the two and is often called the "stability boundary." If these solutions are for an aeroelastic system, the dynamically unstable condition is called "flutter" and the stability boundary corresponding to simple harmonic motion is called the "flutter boundary."

Recall from Eq. (5.13) that the total displacement is a sum of all modal contributions. It is therefore necessary to consider all possible combinations of Γ_k and Ω_k, where Γ_k can be < 0, $= 0$, or > 0 and Ω_k can be $= 0$ or $\neq 0$. The corresponding

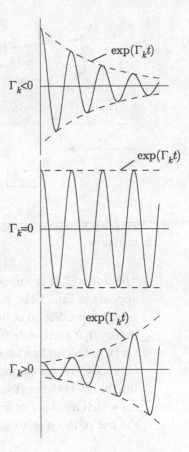

Figure 5.1. Behavior of typical mode amplitude when $\Omega_k \neq 0$

type of motion and stability characteristics are indicated in Table 5.1 for various combinations of Γ_k and Ω_k. Although the primary concern here is in regard to the dynamic instability referred to as flutter, for which $\Omega_k \neq 0$, Table 5.1 shows that the generalized equations of motion also can provide solutions to the static-aeroelastic problem of divergence. This phenomenon is indicated by the unstable condition for $\Omega_k = 0$, and the divergence boundary occurs when $\Gamma_k = \Omega_k = 0$.

In many published works on flutter analysis, the method outlined in this section based on determination of stability from complex eigenvalues is known as the

Table 5.1. *Types of motion and stability characteristics for various values of Γ_k and Ω_k*

Γ_k	Ω_k	Type of motion	Stability characteristic
< 0	$\neq 0$	Convergent Oscillations	Stable
$= 0$	$\neq 0$	Simple Harmonic	Stability Boundary
> 0	$\neq 0$	Divergent Oscillations	Unstable
< 0	$= 0$	Continuous Convergence	Stable
$= 0$	$= 0$	Time Independent	Stability Boundary
> 0	$= 0$	Continuous Divergence	Unstable

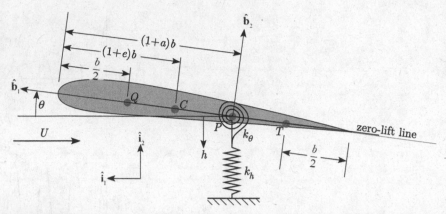

Figure 5.2. Schematic showing geometry of the wing section with pitch and plunge spring restraints

"p method." It is named for the dimensionless complex eigenvalue $p = bv/U$ that appears in Eq. (5.11); p is frequently termed a "reduced eigenvalue." To provide an accurate prediction of flutter characteristics, the p method must use an aerodynamic theory that accurately represents the loads induced by transient motion of the lifting surface. Depending on the theory, augmented aerodynamic states may or may not be necessary; for example, the theory outlined in Section 5.5.2 uses them, whereas the theory in the next section does not; rather it uses the simplest steady-flow theory for which no claim of accuracy is made. The sole purpose of doing so is to illustrate the use of the p method to analyze a simple configuration.

5.2 Aeroelastic Analysis of a Typical Section

In this section, we demonstrate the flutter analysis of a linear aeroelastic system. To do this, a simple model is needed. In the older literature on aeroelasticity, flutter analyses often were performed using simple, spring-restrained, rigid-wing models such as the one shown in Fig. 5.2. These were called "typical-section models" and are still appealing because of their physical simplicity. This configuration could represent the case of a rigid, two-dimensional wind-tunnel model that is elastically mounted in a wind-tunnel test section, or it could correspond to a typical airfoil section along a finite wing. In the latter case, the discrete springs would reflect the wing structural bending and torsional stiffnesses, and the reference point would represent the elastic axis.

Of interest in such models are points P, C, Q, and T, which refer, respectively, to the reference point (i.e., where the plunge displacement h is measured), the center of mass, the aerodynamic center (i.e., presumed to be the quarter-chord in subsonic thin-airfoil theory), and the three-quarter-chord (i.e., an important chordwise location in thin-airfoil theory). The dimensionless parameters e and a (i.e., $-1 \le e \le 1$ and $-1 \le a \le 1$) determine the locations of the points C and P: when these parameters are zero, the points lie on the mid-chord, and when they are positive (negative), the points lie toward the trailing (leading) edge. In the literature, the chordwise offset of the center of mass from the reference point often appears in the equations of motion. It is typically made dimensionless by the airfoil semi-chord b and denoted by $x_\theta = e - a$. This so-called static-unbalance parameter is positive when the center of mass is toward the trailing edge from the reference point. The rigid plunging and pitching of the model is restrained by light, linear springs with spring constants k_h and k_θ.

It is convenient to formulate the equations of motion from Lagrange's equations. To do this, we need kinetic and potential energies, as well as the generalized forces resulting from aerodynamic loading. We immediately can write the potential energy as

$$P = \frac{1}{2}k_h h^2 + \frac{1}{2}k_\theta \theta^2 \tag{5.15}$$

To deduce the kinetic energy, we need the velocity of the mass center C, which can be found as

$$\mathbf{v}_C = \mathbf{v}_P + \dot\theta \hat{\mathbf{b}}_3 \times b\left[(1+a) - (1+e)\right]\hat{\mathbf{b}}_1 \tag{5.16}$$

where the inertial velocity of the reference point P is

$$\mathbf{v}_P = -\dot h \hat{\mathbf{i}}_2 \tag{5.17}$$

and thus

$$\mathbf{v}_C = -\dot h \hat{\mathbf{i}}_2 + b\dot\theta(a - e)\hat{\mathbf{b}}_2 \tag{5.18}$$

The kinetic energy then is given by

$$K = \frac{1}{2}m\mathbf{v}_C \cdot \mathbf{v}_C + \frac{1}{2}I_C \dot\theta^2 \tag{5.19}$$

where I_C is the moment of inertia about C. By virtue of the relationship between $\hat{\mathbf{b}}_2$ and the inertially fixed unit vectors $\hat{\mathbf{i}}_1$ and $\hat{\mathbf{i}}_2$, assuming θ to be small, we find that

$$\begin{aligned}
K &= \frac{1}{2}m\left(\dot h^2 + b^2 x_\theta^2 \dot\theta^2 + 2bx_\theta \dot h \dot\theta\right) + \frac{1}{2}I_C \dot\theta^2 \\
&= \frac{1}{2}m\left(\dot h^2 + 2bx_\theta \dot h \dot\theta\right) + \frac{1}{2}I_P \dot\theta^2
\end{aligned} \tag{5.20}$$

where $I_P = I_C + mb^2 x_\theta^2$.

The generalized forces associated with the degrees of freedom h and θ are derived easily from the work done by the aerodynamic lift through a virtual displacement of the point Q and by the aerodynamic pitching moment about Q through a virtual rotation of the model. The velocity of Q is

$$\mathbf{v}_Q = -\dot h \hat{\mathbf{i}}_2 + b\dot\theta\left(\frac{1}{2} + a\right)\hat{\mathbf{b}}_2 \tag{5.21}$$

The virtual displacement of the point Q can be obtained simply by replacing the dot over each unknown in Eq. (5.21) with a δ in front of it; that is

$$\delta \mathbf{p}_Q = -\delta h \hat{\mathbf{i}}_2 + b\,\delta\theta\left(\frac{1}{2} + a\right)\hat{\mathbf{b}}_2 \tag{5.22}$$

where $\delta\mathbf{p}_Q$ is the virtual displacement at Q. The angular velocity of the wing is $\dot\theta\hat{\mathbf{b}}_3$; therefore, the virtual rotation of the wing is simply $\delta\theta\hat{\mathbf{b}}_3$. Hence, the virtual work of the aerodynamic forces is

$$\overline{\delta W} = L\left[-\delta h + b\left(\frac{1}{2}+a\right)\delta\theta\right] + M_{\frac{1}{4}}\delta\theta \tag{5.23}$$

and the generalized forces become

$$Q_h = -L$$
$$Q_\theta = M_{\frac{1}{4}} + b\left(\frac{1}{2}+a\right)L \tag{5.24}$$

It is clear that the generalized force associated with h is the negative of the lift, whereas the generalized force associated with θ is the pitching moment about the reference point P.

Lagrange's equations (see the Appendix, Eqs. A.35) are specialized here for the case in which the kinetic energy K depends on only $\dot q_1, \dot q_2, \ldots$; therefore

$$\frac{d}{dt}\left(\frac{\partial K}{\partial \dot q_i}\right) + \frac{\partial P}{\partial q_i} = Q_i \qquad (i = 1, 2, \ldots, n) \tag{5.25}$$

Here, $n = 2$, $q_1 = h$, and $q_2 = \theta$ and the equations of motion become

$$m\left(\ddot h + bx_\theta\ddot\theta\right) + k_h h = -L$$
$$I_P\ddot\theta + mbx_\theta\ddot h + k_\theta\theta = M_{\frac{1}{4}} + b\left(\frac{1}{2}+a\right)L \tag{5.26}$$

For the aerodynamics, the steady-flow theory used previously gives

$$L = 2\pi\rho_\infty bU^2\theta$$
$$M_{\frac{1}{4}} = 0 \tag{5.27}$$

where, in accord with thin-airfoil theory, we have taken the lift–curve slope to be 2π. Assuming this representation to be adequate for now, we can apply the p method because the aerodynamic loads are specified for arbitrary motion. (We subsequently consider more sophisticated aerodynamic theories.)

To simplify the notation, we introduce the uncoupled, natural frequencies at zero airspeed, defined by

$$\omega_h = \sqrt{\frac{k_h}{m}} \qquad \omega_\theta = \sqrt{\frac{k_\theta}{I_P}} \tag{5.28}$$

Substituting Eqs. (5.27) into Eqs. (5.26), using the definitions in Eqs. (5.28), and rearranging the equations of motion into matrix form, we obtain

$$\begin{bmatrix} mb^2 & mb^2 x_\theta \\ mb^2 x_\theta & I_P \end{bmatrix}\begin{Bmatrix} \frac{\ddot h}{b} \\ \ddot\theta \end{Bmatrix} + \begin{bmatrix} mb^2\omega_h^2 & 2\pi\rho_\infty b^2 U^2 \\ 0 & I_P\omega_\theta^2 - 2\left(\frac{1}{2}+a\right)\pi\rho_\infty b^2 U^2 \end{bmatrix}\begin{Bmatrix} \frac{h}{b} \\ \theta \end{Bmatrix} = \begin{Bmatrix} 0 \\ 0 \end{Bmatrix}$$
$$\tag{5.29}$$

Note that the first equation was multiplied through by b and the variable h was divided by b to make every term in both equations have the same units. Following the p method as outlined previously, we now make the substitutions $h = \bar{h} \exp(\nu t)$ and $\theta = \bar{\theta} \exp(\nu t)$, which yields

$$\begin{bmatrix} mb^2\nu^2 + mb^2\omega_h^2 & mb^2\nu^2 x_\theta + 2\pi\rho_\infty b^2 U^2 \\ mb^2\nu^2 x_\theta & I_P\omega_\theta^2 + I_P\nu^2 - 2(a + \frac{1}{2})\pi\rho_\infty b^2 U^2 \end{bmatrix} \begin{Bmatrix} \frac{\bar{h}}{b} \\ \bar{\theta} \end{Bmatrix} = \begin{Bmatrix} 0 \\ 0 \end{Bmatrix}. \tag{5.30}$$

Although this eigenvalue problem can be solved as it is written, it is more convenient to introduce dimensionless variables to further simplify the problem. To this end, we first let $\nu = pU/b$, where p is the unknown dimensionless, complex eigenvalue; divide all equations by mU^2; and finally introduce the dimensionless parameters

$$r^2 = \frac{I_P}{mb^2} \qquad \sigma = \frac{\omega_h}{\omega_\theta}$$
$$\mu = \frac{m}{\rho_\infty \pi b^2} \qquad V = \frac{U}{b\omega_\theta} \tag{5.31}$$

Here, r is the dimensionless radius of gyration of the section about the reference point P with $r^2 > x_\theta^2$; σ is the ratio of uncoupled plunge and pitch frequencies; μ is the mass-ratio parameter reflecting the relative importance of the model mass to the mass of the air affected by the model; and V is the dimensionless freestream speed of the air, sometimes called the "reduced velocity." As a result, the equations then simplify to

$$\begin{bmatrix} p^2 + \frac{\sigma^2}{V^2} & x_\theta p^2 + \frac{2}{\mu} \\ x_\theta p^2 & r^2 p^2 + \frac{r^2}{V^2} - \frac{2}{\mu}(a + \frac{1}{2}) \end{bmatrix} \begin{Bmatrix} \frac{\bar{h}}{b} \\ \bar{\theta} \end{Bmatrix} = \begin{Bmatrix} 0 \\ 0 \end{Bmatrix} \tag{5.32}$$

For a nontrivial solution to exist, the determinant of the coefficient matrix must be set equal to zero. There are typically two complex conjugate pairs of roots—for example

$$p_1 = \frac{b\nu_1}{U} = \frac{b}{U}(\Gamma_1 \pm i\Omega_1)$$
$$p_2 = \frac{b\nu_2}{U} = \frac{b}{U}(\Gamma_2 \pm i\Omega_2) \tag{5.33}$$

A more convenient way to present these roots is to multiply them by the reduced velocity V, yielding

$$Vp_1 = \frac{b}{U}(\Gamma_1 \pm i\Omega_1)\frac{U}{b\omega_\theta} = \frac{\Gamma_1}{\omega_\theta} \pm i\frac{\Omega_1}{\omega_\theta}$$
$$Vp_2 = \frac{b}{U}(\Gamma_2 \pm i\Omega_2)\frac{U}{b\omega_\theta} = \frac{\Gamma_2}{\omega_\theta} \pm i\frac{\Omega_2}{\omega_\theta} \tag{5.34}$$

This way, they are now tied to a specified system parameter ω_θ instead of the varying speed U.

For a given configuration and altitude, we must look at the behavior of the complex roots as functions of V and find the smallest value of V to give divergent

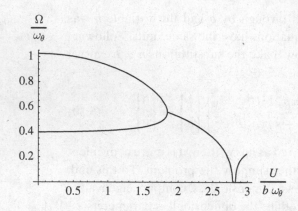

Figure 5.3. Plot of the modal frequency versus V for $a = -1/5$, $e = -1/10$, $\mu = 20$, $r^2 = 6/25$, and $\sigma = 2/5$ (steady-flow theory)

oscillations in accordance with Table 5.1. That value is $V_F = U_F/(b\omega_\theta)$, where U_F is the flutter speed.

We may find the divergence speed by setting $p = 0$ in Eq. (5.32), which leads to setting the coefficient of $\bar{\theta}$ in the $\bar{\theta}$ equation equal to zero and solving the resulting expression for V. This value is the dimensionless divergence speed V_D, given by

$$V_D = \frac{U_D}{b\omega_\theta} = r\sqrt{\frac{\mu}{1 + 2a}} \tag{5.35}$$

This is the same answer that we would obtain with analyses similar to those presented in Chapter 4.

For looking at flutter, we consider a specific configuration defined by $a = -1/5$, $e = -1/10$, $\mu = 20$, $r^2 = 6/25$, and $\sigma = 2/5$. The divergence speed for this configuration is $V_D = 2.828$ (or $U_D = 2.828\, b\,\omega_\theta$). Plots of the imaginary and real parts of the roots versus V are shown in Figs. 5.3 and 5.4, respectively. The negative of Γ is the modal damping and Ω is the modal frequency. We consider first the imaginary parts, Ω, as shown in Fig. 5.3. When $V = 0$, we expect the two dimensionless frequencies to be near unity and σ for pitching and plunging oscillations, respectively. Even at $V = 0$, these modes are lightly coupled because of the nonzero off-diagonal terms proportional to x_θ in the mass matrix. As V increases, the frequencies start to

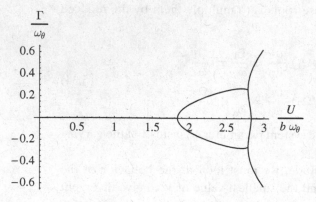

Figure 5.4. Plot of the modal damping versus V for $a = -1/5$, $e = -1/10$, $\mu = 20$, $r^2 = 6/25$, and $\sigma = 2/5$ (steady-flow theory)

approach one another, and their respective mode shapes exhibit increasing coupling between plunge and pitch. Flutter occurs when the two modal frequencies coalesce, at which point the roots become complex conjugate pairs. At this condition, both modes are highly coupled pitch–plunge oscillations. The dimensionless flutter speed is $V_F = U_F/(b\omega_\theta) = 1.843$ and the flutter frequency is $\Omega_F/\omega_\theta = 0.5568$. The real parts, Γ, are shown in Fig. 5.4 and remain zero until flutter occurs. When flutter occurs, the real part of one of the roots is positive and the other is negative.

Comparing results from this analysis with experimental data, we find that a few elements of realism are at least qualitatively captured. For example, the analysis predicts that flutter occurs at a value of $V = V_F < V_D$, which is correct for the specified configuration. Furthermore, it shows a coalescence of the pitching and plunging frequencies as V approaches V_F, which is not only correct for the specified configuration but also is frequently observed in connection with flutter analysis. However, the previous analysis is deficient in its ability to accurately predict flutter speed. Moreover, the damping of all modes below the flutter speed is predicted to be zero, which is known to be incorrect. Finally, the steady-flow theory exhibits a coalescence characterized by the two roots being exactly equal to one another at the point of flutter. This condition is not met at all in data obtained from experiments and flight testing.

These deficiencies in predictive capability stem from deficiencies in the aerodynamic theory. The steady-flow aerodynamic theory of Chapter 4 was used. Although this aerodynamic theory has obvious deficiencies (e.g., linearity and two-dimensionality), a most significant deficiency concerning flutter analysis is that it neglects unsteady effects. To obtain an accurate prediction of flutter speed, it is necessary to include unsteadiness in the aerodynamic theory; this demands a more sophisticated aerodynamic theory.

Unfortunately, development of unsteady-aerodynamic theories is no small undertaking. Unsteady-aerodynamic theories can be developed most simply when simple harmonic motion is assumed a priori. Although such limited theories cannot be used in the p method of flutter analysis described in Section 5.1, they can be used in classical flutter analysis, described in the next section. As will be shown, classical flutter analysis can predict the flutter speed and flutter frequency, but it cannot predict values of modal damping and frequency away from the flutter condition. To obtain a reasonable sense of modal damping and frequencies at points other than the flutter condition, two approximate schemes are discussed in Section 5.4.

If these approximations turn out to be inadequate for predicting modal damping and frequencies, we have no choice but to carry out a flutter analysis that does not assume simple harmonic motion, which in turn requires a still more powerful aerodynamic theory. One such approach that fits easily into the framework of Section 5.1 is the finite-state theory of Peters et al. (1995). Such a theory not only facilitates the calculation of subcritical eigenvalues; because it is a time-domain model, it also can be used in control design.

Hence, in the following sections, we first look at classical flutter analysis and the approximate techniques associated therewith and then turn to a more detailed discussion of unsteady aerodynamics, including one theory that assumes simple harmonic motion (i.e., the Theodorsen theory) and one that does not (i.e., the Peters finite-state theory).

5.3 Classical Flutter Analysis

Until at least the late 1970s, the aircraft industry performed most lifting-surface flutter analyses based on what is commonly called "classical flutter analysis" based on the flutter determinant. The objective of such an analysis is to determine the flight conditions that correspond to the flutter boundary. It was previously noted that the flutter boundary corresponds to conditions for which one of the modes of motion has a simple harmonic time dependency. Because this is considered to be a stability boundary, it is implied that all modes of motion are convergent (i.e., stable) for less critical flight conditions (i.e., lower airspeed). Moreover, all modes other than the critical one are convergent at the flutter boundary.

The method of analysis is not based on solving the generalized equations of motion as described in Section 5.1. Rather, it is presumed that the solution involves simple harmonic motion. With such a solution specified, the equations of motion are then solved for the flight condition(s) that yields such a solution. Whereas in the p method we determine the eigenvalues for a set flight condition—the real parts of which provide the modal damping—it is apparent that classical flutter analysis cannot provide the modal damping for an arbitrary flight condition. Thus, it cannot provide any definitive measure of flutter stability other than the location of the stability boundary. Although this is the primary weakness of such a method, its primary strength is that it needs only the unsteady airloads for simple harmonic motion of the surface, which for a given level of accuracy are derived more easily than those for arbitrary motion.

To illustrate classical flutter analysis, it is necessary to consider an appropriate representation of unsteady airloads for simple harmonic motion of a lifting surface. Because these oscillatory motions are relatively small in amplitude, it is sufficient to use a linear-aerodynamic theory for the computation of these loads. These aero-dynamic theories usually are based on linear potential-flow theory for thin airfoils, which presumes that the motion and thickness of the wing structure create a small disturbance in the flow field and that perturbations in the flow velocity are small relative to the freestream speed. For purposes of demonstration, it suffices to reconsider the typical section of a two-dimensional lifting surface that is experiencing simultaneous translational and rotational motions, as illustrated in Fig. 5.2. The motion is simple harmonic; thus, h and θ are represented as

$$h(t) = \overline{h}\exp(i\omega t)$$
$$\theta(t) = \overline{\theta}\exp(i\omega t)$$

(5.36)

where ω is the frequency of the motion. Although the h and θ motions are of the same frequency, they are not necessarily in phase. This can be taken into account mathematically by representing the amplitude $\overline{\theta}$ as a real number and $\overline{h}$ as a complex number. Because a linear aerodynamic theory is to be used, the resulting lift, L, and the pitching moment about P, denoted by M, where

$$M = M_{\frac{1}{4}} + b\left(\frac{1}{2} + a\right)L \tag{5.37}$$

also are simple harmonic with frequency ω, so that

$$L(t) = \overline{L}\exp(i\omega t)$$
$$M(t) = \overline{M}\exp(i\omega t) \tag{5.38}$$

The amplitudes of these airloads can be computed as complex linear functions of the amplitudes of motion as

$$\overline{L} = -\pi\rho_\infty b^3\omega^2\left[\ell_h(k, M_\infty)\frac{\overline{h}}{b} + \ell_\theta(k, M_\infty)\overline{\theta}\right]$$
$$\overline{M} = \pi\rho_\infty b^4\omega^2\left[m_h(k, M_\infty)\frac{\overline{h}}{b} + m_\theta(k, M_\infty)\overline{\theta}\right] \tag{5.39}$$

Here, the freestream air density is represented as ρ_∞ and the four complex functions contained in the square brackets represent the dimensionless aerodynamic coefficients for the lift and moment resulting from plunging and pitching. These coefficients in general, are, functions of the two parameters k and M_∞, where

$$k = \frac{b\omega}{U} \quad \text{(reduced frequency)}$$
$$M_\infty = \frac{U}{c_\infty} \quad \text{(freestream Mach number)} \tag{5.40}$$

As in the case of steady airloads, compressibility effects are reflected here by the dependence of the coefficients on M_∞. The reduced frequency k is unique to unsteady flows. This dimensionless frequency parameter is a measure of the unsteadiness of the flow and normally has a value between zero and unity for conventional flight vehicles. Also note that for any specified values of k and M_∞, each coefficient can be written as a complex number. As in the case of $\overline{h}$ relative to $\overline{\theta}$, the fact that lift and pitching moment are complex quantities reflects their phase relationships relative to the pitch angle (where we can regard $\overline{\theta}$ as a real number, for convenience). The speed at which flutter occurs corresponds to specific values of k and M_∞ and must be found by iteration. Examples of how this process can be carried out for one- and two-degree-of-freedom systems are given in the following subsections.

5.3.1 One-Degree-of-Freedom Flutter

To illustrate the application of classical flutter analysis, a simple configuration is treated first. This example is a one-degree-of-freedom aeroelastic system consisting of a rigid two-dimensional wing that is permitted to rotate in pitch about a specified

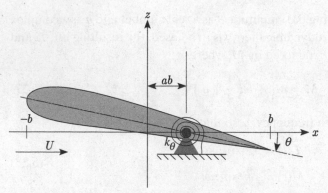

Figure 5.5. Schematic of the airfoil of a two-dimensional wing that is spring-restrained in pitch

reference point. This is a special case of the typical-section configuration shown in Fig. 5.2 for which the plunge degree of freedom is equal to zero, as depicted in Fig. 5.5. The system equations of motion reduce to one equation that can be written as

$$I_P\ddot{\theta} + k_\theta\theta = M \tag{5.41}$$

To be consistent with classical flutter analysis, the motion of the system is presumed to be simple harmonic as

$$\theta = \bar{\theta}\exp(i\omega t) \tag{5.42}$$

The aerodynamic pitching moment, M, in the equation of motion is in response to this simple harmonic pitching displacement. As previously discussed, this airload can be described by

$$M = \overline{M}\exp(i\omega t) \tag{5.43}$$

where

$$\overline{M} = \pi\rho_\infty b^4\omega^2 m_\theta(k, M_\infty)\bar{\theta} \tag{5.44}$$

Substituting these simple harmonic functions into the equation of motion yields an algebraic relationship between the coefficients of $\bar{\theta}$ as

$$k_\theta - \omega^2 I_P = \pi\rho_\infty b^4\omega^2 m_\theta(k, M_\infty) \tag{5.45}$$

Introducing the natural frequency of the system at zero airspeed

$$\omega_\theta = \sqrt{\frac{k_\theta}{I_P}} \tag{5.46}$$

and rearranging the algebraic relationship, we obtain the final equation to be solved for the flight condition at the flutter boundary as

$$\frac{I_P}{\pi\rho_\infty b^4}\left[1 - \left(\frac{\omega_\theta}{\omega}\right)^2\right] + m_\theta(k, M_\infty) = 0 \tag{5.47}$$

To solve this equation, it is presumed that the configuration parameters I_P, ω_θ, and b are known. The unknown parameters that describe the motion and flight condition are ω, ρ_∞, k, and M_∞. These four unknowns must be determined from the single algebraic equation, Eq. (5.47). Because the aerodynamic coefficient, $m_\theta(k, M_\infty)$, is complex, it can be written as

$$m_\theta(k, M_\infty) = \Re[m_\theta(k, M_\infty)] + i\Im[m_\theta(k, M_\infty)] \qquad (5.48)$$

As a consequence, both the real and imaginary parts of the algebraic relationship must be zero, thus providing two real equations to determine the four unknowns. Therefore, two of the unknown parameters should be specified. A fixed altitude is chosen that specifies the freestream atmospheric density, ρ_∞. The second parameter to be fixed is the Mach number, which can be given a temporary value of zero. This, of course, implies that the flow is incompressible and the aerodynamic-moment coefficient is then only a function of the reduced frequency. The governing algebraic equation now can be written as

$$\frac{I_P}{\pi \rho_\infty b^4}\left[1 - \left(\frac{\omega_\theta}{\omega}\right)^2\right] + \Re[m_\theta(k, 0)] + i\Im[m_\theta(k, 0)] = 0 \qquad (5.49)$$

Equating the imaginary part of the left-hand side to zero gives a relationship that can be solved for the reduced frequency, k_F, at the flutter boundary; that is

$$\Im[m_\theta(k_F, 0)] = 0 \qquad (5.50)$$

With k_F known, $\Re[m_\theta(k_F, 0)]$ can be numerically evaluated. Equating the real part of the left-hand side to zero now enables the frequency, ω_F, to be determined from

$$\left(\frac{\omega_\theta}{\omega_F}\right)^2 = 1 + \frac{\pi \rho_\infty b^4 \Re[m_\theta(k_F, 0)]}{I_P} \qquad (5.51)$$

Now that k_F and ω_F have been determined, it is possible to compute the flutter speed as

$$U_F = \frac{b\omega_F}{k_F} \qquad (5.52)$$

The flutter speed determined by the previous procedure corresponds to the originally specified altitude and is based on an incompressible representation of the airloads. After this speed has been determined, the speed of sound, c_∞, at the specified altitude can be used to find the flutter Mach number as

$$M_F = \frac{U_F}{c_\infty} \qquad (5.53)$$

If this flutter Mach number is sufficiently small to justify the use of incompressible aerodynamic coefficients, then the altitude–speed combination obtained is a point on the flutter boundary. If the flutter Mach number is too high to validate the incompressible approximation, then the entire procedure should be repeated using aerodynamic coefficients that are based on the initially computed flutter Mach number. Using the standard atmospheric model, which relates density and the speed of sound, this iterative scheme converges to a flight condition on the flutter boundary.

5.3.2 Two-Degree-of-Freedom Flutter

The analysis of multi-degree-of-freedom systems for determination of the flutter boundary can be demonstrated adequately by the simple two-degree-of-freedom configuration in Fig. 5.2. The equations of motion, already derived as Eqs. (5.26), are repeated here as follows:

$$m\left(\ddot{h} + bx_\theta\ddot{\theta}\right) + k_h h = -L$$
$$I_P\ddot{\theta} + mbx_\theta\ddot{h} + k_\theta\theta = M \tag{5.54}$$

where, as before

$$M = M_{\frac{1}{4}} + b\left(\frac{1}{2} + a\right)L \tag{5.55}$$

The next step in classical flutter analysis is to presume that the motion is simple harmonic as represented by

$$h = \overline{h}\exp(i\omega t)$$
$$\theta = \overline{\theta}\exp(i\omega t) \tag{5.56}$$

The corresponding lift and moment can be written as

$$L = \overline{L}\exp(i\omega t)$$
$$M = \overline{M}\exp(i\omega t) \tag{5.57}$$

Substituting these time-dependent functions into the equations of motion, we obtain a pair of algebraic equations for the amplitudes of h and θ in the form

$$-\omega^2 m\overline{h} - \omega^2 mbx_\theta\overline{\theta} + m\omega_h^2\overline{h} = -\overline{L}$$
$$-\omega^2 mbx_\theta\overline{h} - \omega^2 I_P\overline{\theta} + I_P\omega_\theta^2\overline{\theta} = \overline{M} \tag{5.58}$$

where we recall that

$$\overline{L} = -\pi\rho_\infty b^3\omega^2\left[\ell_h\left(k, M_\infty\right)\frac{\overline{h}}{b} + \ell_\theta\left(k, M_\infty\right)\overline{\theta}\right]$$

$$\overline{M} = \pi\rho_\infty b^4\omega^2\left[m_h\left(k, M_\infty\right)\frac{\overline{h}}{b} + m_\theta\left(k, M_\infty\right)\overline{\theta}\right] \tag{5.59}$$

Substituting these lift and moment amplitudes into Eqs. (5.58) and then rearranging, we obtain a pair of homogeneous, linear, algebraic equations for $\overline{h}$ and $\overline{\theta}$, given by

$$\left\{\frac{m}{\pi\rho_\infty b^2}\left[1 - \left(\frac{\omega_h}{\omega}\right)^2\right] + \ell_h\left(k, M_\infty\right)\right\}\frac{\overline{h}}{b} + \left[\frac{mx_\theta}{\pi\rho_\infty b^2} + \ell_\theta\left(k, M_\infty\right)\right]\overline{\theta} = 0$$

$$\left[\frac{mx_\theta}{\pi\rho_\infty b^2} + m_h\left(k, M_\infty\right)\right]\frac{\overline{h}}{b} + \left\{\frac{I_P}{\pi\rho_\infty b^4}\left[1 - \left(\frac{\omega_\theta}{\omega}\right)^2\right] + m_\theta\left(k, M_\infty\right)\right\}\overline{\theta} = 0 \tag{5.60}$$

The coefficients in these equations that involve the inertia terms are symbolically simplified by defining the dimensionless parameters used previously; namely

$$\mu = \frac{m}{\pi \rho_\infty b^2} \quad \text{(mass ratio)}$$

$$r = \sqrt{\frac{I_P}{mb^2}} \quad \text{(mass radius of gyration about } P\text{)}$$

(5.61)

Using these parameters allows us to rewrite the previous two homogeneous equations in a simpler way:

$$\left\{ \mu \left[1 - \left(\frac{\omega_h}{\omega} \right)^2 \right] + \ell_h \right\} \frac{\overline{h}}{b} + (\mu x_\theta + \ell_\theta) \overline{\theta} = 0$$

$$(\mu x_\theta + m_h) \frac{\overline{h}}{b} + \left\{ \mu r^2 \left[1 - \left(\frac{\omega_\theta}{\omega} \right)^2 \right] + m_\theta \right\} \overline{\theta} = 0$$

(5.62)

The third step in the flutter analysis is to solve these algebraic equations for the flight condition(s) for which the presumed simple harmonic motion is valid. This result corresponds to the flutter boundary. If it is presumed that the configuration parameters m, e, a, I_P, ω_h, ω_θ, and b are known, then the unknown quantities $\overline{h}$, $\overline{\theta}$, ω, ρ_∞, M_∞, and k describe the motion and flight condition. Because Eqs. (5.62) are linear and homogeneous in $\overline{h}/b$ and $\overline{\theta}$, the determinant of their coefficients must be zero for a nontrivial solution for the motion to exist. This condition can be written as

$$\begin{vmatrix} \mu \left[1 - \sigma^2 \left(\frac{\omega_\theta}{\omega} \right)^2 \right] + \ell_h & \mu x_\theta + \ell_\theta \\ \mu x_\theta + m_h & \mu r^2 \left[1 - \left(\frac{\omega_\theta}{\omega} \right)^2 \right] + m_\theta \end{vmatrix} = 0$$

(5.63)

The determinant in this relationship is called the "flutter determinant." Note that the parameter $\sigma = \omega_h / \omega_\theta$ was introduced so that a common term that is explicit in ω is available—namely, ω_θ / ω. Thus, expansion of the determinant yields a quadratic polynomial in the unknown $\lambda = (\omega_\theta / \omega)^2$.

To complete the solution for the flight condition at the flutter boundary, it must be recognized that four unknowns remain: ω_θ / ω, $\mu = m/(\pi \rho_\infty b^2)$, M_∞, and $k = b\omega / U$. The one equation available for their solution is the second-degree polynomial characteristic equation from setting the determinant equal to zero. However, because the aerodynamic coefficients are complex quantities, this complex equation represents two real equations, wherein both the real and imaginary parts must be identically zero for a solution to be obtained. This means that two of the four unknowns must be specified. A procedure to solve for and map the flutter boundary is outlined as follows:

1. Specify an altitude, which fixes the parameter μ.
2. Specify an initial guess for M_∞ of, say, zero.

3. Recalling that setting the flutter determinant equal to zero yields a quadratic equation in λ, use a root-finding application[1] to find the value of k at which the imaginary part of one of the two roots for λ vanishes, which is k_F. This can be carried out easily with computerized symbolic manipulation software such as Mathematica™ or Maple.™

4. Set $\omega_\theta/\omega_F = \sqrt{\lambda(k_F)}$ using the root for which $\lambda(k_F)$ is real.

5. Determine $U_F = b\omega_F/k_F$ and $M_{\infty_F} = U_F/c_\infty$.

6. Repeat steps 3–5 with the value of M_{∞_F} obtained in step 5 until converged values are obtained for M_{∞_F}, k_F, and U_F for flutter at a given μ.

7. Repeat the entire procedure for various values of μ (i.e., an indication of the altitude for a given aircraft) to determine the flutter boundary in terms of, say, altitude versus M_{∞_F}, k_F, and U_F.

5.4 Engineering Solutions for Flutter

It was noted in the preceding section that the presumption of simple harmonic motion in classical flutter analysis has both advantages and disadvantages. The prime argument for specification of simple harmonic time dependency is, of course, its correspondence to the stability boundary. Identification of the flight conditions along this boundary requires the execution of a tedious, iterative process such as the one outlined in Section 5.3. This type of solution can be attributed to Theodorsen (1934), who presented the first comprehensive flutter analysis with his development of the unsteady airloads on a two-dimensional wing in incompressible potential flow.

Although unsteady-aerodynamics analyses for simple harmonic motion are not simple to formulate and execute, they are far more tractable than those for oscillatory motions with varying amplitude. Since the work of Theodorsen, numerous unsteady-aerodynamic formulations have been developed for simple harmonic motion of lifting surfaces. These techniques have proven to be adequate for compressible flows in both the subsonic and supersonic regimes. They also have been developed for three-dimensional surfaces and, in some cases, with surface-to-surface interaction. This availability of relatively accurate unsteady-aerodynamic theories for simple harmonic motion was the stimulus for further development of flutter analyses beyond that of the classical flutter analysis described in Section 5.3.

---

[1] If one does not have ready access to a root-finding application, this step may be replaced by the following four steps:

   (a) Specify a set of trial k values—say, from 0.001 to 1.0.

   (b) For each value of k (and the specified value of M_∞), calculate the functions ℓ_h, ℓ_θ, m_h, and m_θ.

   (c) Solve the flutter determinant, which is a quadratic equation with complex coefficients, for the values of $\lambda = (\omega_\theta/\omega)^2$ that correspond to each of the selected values of k. Note that these roots are complex in general, the real part an approximation of $(\omega_\theta/\omega)^2$ and the imaginary part related to the damping of the mode.

   (d) Interpolate to find the value of k at which the imaginary part of one of the roots becomes zero. This can be done approximately by plotting the imaginary parts of both roots versus k, so that the value of k at which one of the imaginary parts crosses the zero axis can be determined. This value of k is an approximation of k_F, making the value of λ real when $k = k_F$.

There are two other important considerations of practicing engineers. The first is to obtain an understanding of the margin of stability at flight conditions in the vicinity of the flutter boundary. The second—and possibly the more important—is to obtain an understanding of the physical mechanism that causes the instability. With this information, engineers can propose design variations that may alleviate or even eliminate the instability. When a suitable unsteady-aerodynamic theory is available, the p method can address these considerations. In this section, we examine alternative ways that engineers have addressed these problems when unsteady-aerodynamic theories that assume simple harmonic motion must be used.

5.4.1 The k Method

Subsequent to Theodorsen's analysis of the flutter problem, numerous schemes were devised to extract the roots of the "flutter determinant" and thereby identify the stability boundary. Scanlan and Rosenbaum (1951) presented a brief overview of these techniques as they were offered during the 1940s. It was fairly common to include in the flutter analysis a parameter that simulated the effect of structural damping. Observations at that time indicated that the energy removed per cycle during a simple harmonic oscillation was nearly proportional to the square of the amplitude but independent of the frequency. This behavior can be characterized by a damping force that is proportional to the displacement but in phase with the velocity.

To incorporate this form of structural damping into the analysis of Section 5.3.2, Eqs. (5.54) can be written as

$$m\left(\ddot{h} + bx_\theta\ddot{\theta}\right) + k_h h = -L + D_h$$
$$I_P\ddot{\theta} + mbx_\theta\ddot{h} + k_\theta\theta = M + D_\theta \tag{5.64}$$

where the dissipative structural damping terms are

$$D_h = \overline{D}_h \exp(i\omega t)$$
$$= -i g_h m\omega_h^2 \overline{h} \exp(i\omega t)$$
$$D_\theta = \overline{D}_\theta \exp(i\omega t) \tag{5.65}$$
$$= -i g_\theta I_P \omega_\theta^2 \overline{\theta} \exp(i\omega t)$$

Proceeding as before, Eqs. (5.62) become

$$\left\{\mu\left[1 - \left(\frac{\omega_h}{\omega}\right)^2 (1 + ig_h)\right] + \ell_h\right\}\frac{\overline{h}}{b} + (\mu x_\theta + \ell_\theta)\overline{\theta} = 0$$
$$(\mu x_\theta + m_h)\frac{\overline{h}}{b} + \left\{\mu r^2\left[1 - \left(\frac{\omega_\theta}{\omega}\right)^2 (1 + ig_\theta)\right] + m_\theta\right\}\overline{\theta} = 0 \tag{5.66}$$

The damping coefficients g_h and g_θ have representative values from 0.01 to 0.05 depending on the structural configuration. Most early analysts who incorporated this type of structural damping model into their flutter analyses specified the coefficient values a priori with the intention of improving the accuracy of their results.

Scanlan and Rosenbaum (1948) suggested that the damping coefficients be treated as unknown together with ω. In this instance, the subscripts on g can be removed. Writing $\sigma = \omega_h/\omega_\theta$ as before, and introducing

$$Z = \left(\frac{\omega_\theta}{\omega}\right)^2 (1 + ig) \tag{5.67}$$

we obtain the flutter determinant as

$$\begin{vmatrix} \mu\left(1 - \sigma^2 Z\right) + \ell_h & \mu x_\theta + \ell_\theta \\ \mu x_\theta + m_h & \mu r^2\left(1 - Z\right) + m_\theta \end{vmatrix} = 0 \tag{5.68}$$

which is a quadratic equation in Z. The two unknowns of this quadratic equation are complex, denoted by

$$Z_{1,2} = \left(\frac{\omega_\theta}{\omega_{1,2}}\right)^2 (1 + ig_{1,2}) \tag{5.69}$$

The computational strategy for solving Eq. (5.68) proceeds in a manner similar to the one outlined for Eq. (5.63). The primary difference is that the numerical results consist of two pairs of real numbers, (ω_1, g_1) and (ω_2, g_2), which can be plotted versus airspeed U or a suitably normalized value such as $U/(b\omega_\theta)$ or "reduced velocity" $1/k$.

Plots of the damping coefficients g_1 and g_2 versus airspeed can indicate the margin of stability at conditions near the flutter boundary, where g_1 or g_2 is equal to zero. These plots proved to be of such significance that the technique of incorporating the unknown structural damping was initially called the "U-g method." Recalling that the methodology presumes simple harmonic motion throughout, the numerical values of g_1 and g_2 that are obtained for each k can be interpreted only as the required damping coefficients (of the specified form) to achieve simple harmonic motion at frequencies ω_1 and ω_2, respectively. The damping as modeled does not really exist; it was introduced as an artifice to produce the desired motion—truly an artificial structural damping.

The plots of frequency versus airspeed in conjunction with the damping plots can, in many cases, provide an indication of the physical mechanism that leads to the instability. The values of frequency along the $U = 0$ axis correspond to the coupled modes of the original structural dynamic system. As the airspeed increases, the individual behavior or interaction of these roots can indicate the transfer of energy from one mode to another. Such observations could suggest a way to delay the onset of the instability. To confirm identification of the modes of motion for any specified reduced frequency, it is only necessary to substitute the corresponding eigenvalues, ω_i and g_i, into the homogeneous equations of motion to compute the associated eigenvector $(\overline{h}/b, \overline{\theta})$. Because this is a complex number, it can provide the relative magnitude and phase of the original deflections h and θ.

5.4.2 The p-k Method

The k method is still popular in industry largely because of its speed. Although it provides significant advantages to the practicing aeroelastician, it is a mathematically improper formulation. The impropriety of imposing simple harmonic motion with the introduction of artificial damping precipitated many heated discussions through-out the industry. It has been argued that for conditions other than the $g = 0$ case, the frequency and damping characteristics do not correctly represent the system behavior. As a result, design changes that are made based on these characteristics can lead to expensive and potentially dangerous results.

In 1971, Hassig presented definitive numerical results that clearly indicated that the k method of flutter analysis can exhibit an improper coupling among the modes of motion. His presentation utilized a simple form of unsteady aerodynamics in a k method analysis, and then he compared the results with those from a p method analysis. Recall that the p method presented in Section 5.1 is already established as the most accurate solution. Here, we show how both k and p-k methods relate to it.

In the p method, the general solution to the homogeneous modal equations of motion given by Eqs. (5.7) can be written in terms of a dimensionless eigenvalue parameter $p = bv/U$, where

$$\xi_i(t) = \overline{\xi}_i \exp(vt) \tag{5.70}$$

Substitution of this expression into Eqs. (5.7) yields $n + N$ linear, homogeneous equations for the n $\overline{\xi}_i$s and the N $\overline{\lambda}$s given as Eqs. (5.11). After eliminating the $\overline{\lambda}$s using the second of Eqs. (5.11), we may rewrite the first of those equations symbolically as

$$\left[p^2 \lceil M_\cdot \rfloor + \frac{b^2}{U^2} \lceil M_\cdot \rfloor \lceil \omega^2_\cdot \rfloor - \rho_\infty [\mathcal{A}(p)] \right] \{\overline{\xi}\} = 0 \tag{5.71}$$

where $\lceil M_\cdot \rfloor$ and $\lceil \omega^2_\cdot \rfloor$ are diagonal matrices with elements $M_1, M_2, \ldots, M_n$ and $\omega^2_1, \omega^2_2, \ldots, \omega^2_n$, respectively; n is the number of modes; and the unsteady aerody-namics operator matrix $[\mathcal{A}(p)]$ can be expressed in terms of the other matrices in Eqs. (5.11)—namely, in terms of $[a]$, $[b]$, $[c]$, $[d]$, $[A]$, and $[E]$. The complex matrix $[\mathcal{A}(p)]$ is made up of so-called aerodynamic-influence coefficients (AICs). These coefficients are functions of p and possibly of the Mach number, depending on the sophistication of the aerodynamic theory.

If the $\overline{\lambda}$s are actually eliminated, then the problem, in general, cannot be ex-pressed as a standard eigenvalue problem. This is not a serious obstacle, however, because we can always solve Eqs. (5.11) as a standard eigenvalue problem. The purpose here for eliminating the $\overline{\lambda}$s is only to provide a convenient segue into the k and p-k methods and show explicitly the differences among the three methods. The important thing to note is not the procedure used to obtain Eq. (5.71); rather, it is the form of this equation that is most important at this stage. The coefficients $\mathcal{A}(p)$ frequently can be determined or identified in other ways. This equation is the basis for the p method in one of its usual forms. For a nontrivial solution of the generalized coordinate amplitudes, the determinant of the coefficient matrix in

Eq. (5.71) must be zero, so that

$$\left| p^2 [M.] + \frac{b^2}{U^2} [M.][\omega^2.] - \rho_\infty [\mathcal{A}(p)] \right| = 0 \tag{5.72}$$

For a given speed and altitude, this flutter determinant can be solved for p. The result typically yields a set of complex conjugate pairs and real roots, the former represented as

$$p = \gamma k \pm ik \tag{5.73}$$

where k is the reduced frequency of Eqs. (5.40), γ is the rate of decay, given by

$$\gamma = \frac{1}{2\pi} \ln \left(\frac{a_{m+1}}{a_m} \right) \tag{5.74}$$

and where a_m and a_{m+1} represent the amplitudes of successive cycles.

Application of the k method to this modal representation can be readily achieved by letting $p = ik$ in the preceding formulation. This yields a flutter determinant comparable to Eq. (5.72) as

$$\left| -k^2 [M.] + \frac{b^2}{U^2} [M.][\omega^2.] - \rho_\infty [\mathcal{A}(ik)] \right| = 0 \tag{5.75}$$

At selected values of reduced frequency and altitude, Eq. (5.75) can be solved for the complex roots of b^2/U^2, denoted by $\lambda_r + i\lambda_i$. These roots may be interpreted as

$$\lambda_r + i\lambda_i = \left(\frac{b^2}{U^2} \right) (1 + ig) \tag{5.76}$$

where g is the structural damping required to sustain simple harmonic motion. This structural-damping parameter can be related to the rate of decay parameter of the p method as

$$g \cong 2\gamma \tag{5.77}$$

This is a good approximation for small damping as in the case of flight vehicles. The k method is posed easily as a standard eigenvalue problem, which is clear from Eq. (5.75). This alone gives it a significant advantage over the classical flutter-determinant method outlined in Section 5.3.

Another important aspect in making any correlation between the p and k methods is the matter of adequate inclusion of compressibility effects in the unsteady-aerodynamic terms. In the p method, the flutter determinant is solved for selected combinations of speed and altitude. Consequently, the appropriate Mach number can be used for the aerodynamic terms at the outset of the computation. In contrast, the k method preselects combinations of reduced frequency and altitude. As a result of then computing the airspeed as an unknown, λ_r, the Mach number cannot be accurately specified a priori. The result is that an iterative process similar to the one described in Section 5.3 must be conducted to ensure that compressibility effects are adequately incorporated in the k method.

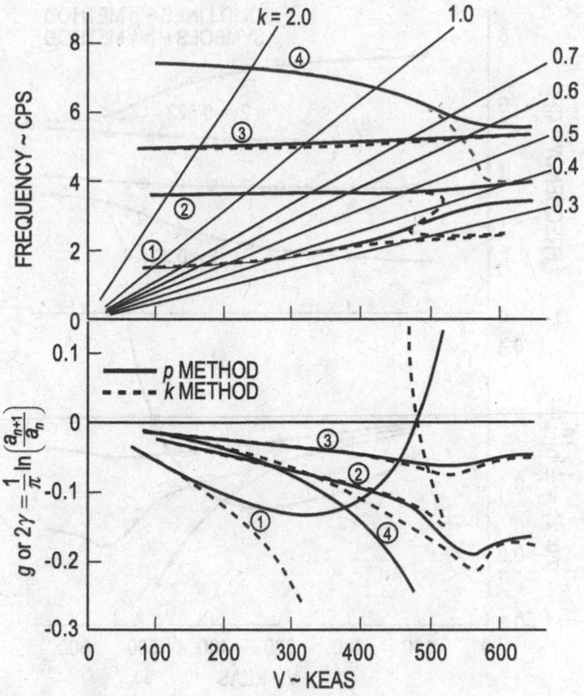

Figure 5.6. Comparison between p and k methods of flutter analysis for a twin-jet transport airplane (from Hassig [1971] Fig. 1, used by permission)

Hassig applied the p and k methods of flutter analysis to a realistic aircraft configuration. By incorporating the same unsteady-aerodynamic representation in each analysis, he was able to make a valid comparison of the results. His observations are typified by Fig. 5.6 (which is his Fig. 1). Note from this figure that not only is the modal coupling wrongly predicted by the k method but also, more important, the wrong mode is predicted to become unstable. The only consistently valid result between the two analyses is that of the flutter speed for which $g = \gamma = 0$. Despite the inconsistent modal coupling exhibited by the k method, it permits the use of simple harmonic modeling of the unsteady aerodynamic terms. As previously mentioned, the accuracy of simple harmonic airload predictions exceeds the accuracy of airload predictions for transient motions. It is for this reason that a compromise between the two models was suggested.

The p-k method is such a compromise. It is based on conducting a p-method type of analysis with the restriction that the unsteady-aerodynamics matrix is for simple harmonic motion. Using an arbitrary value of k in computing $[\mathcal{A}(ik)]$, we find the flutter determinant to be

$$\left| p^2[M] + \frac{b^2}{U^2}[M][\omega^2] - \rho_\infty[\mathcal{A}(ik)] \right| = 0 \qquad (5.78)$$

Given a set of initial guesses for k—say, $k_0 = b\omega_i/U$ for the ith root—this equation can be solved for p. Moreover, it can be posed as a standard eigenvalue problem for p because p appears only in a simple way. The typical result is a set of complex

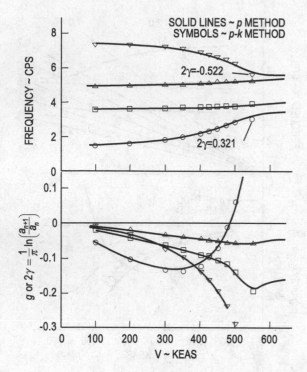

Figure 5.7. Comparison between p and p-k methods of flutter analysis for a twin-jet transport airplane (from Hassig [1971] Fig. 2, used by permission)

conjugate pairs of roots and possibly some real roots. Selecting one of the complex roots and denoting the initial solution as

$$k_1 = |\Im(p)| \qquad \gamma_1 = \frac{\Re(p)}{k_1} \tag{5.79}$$

we can compute $[A(ik_1)]$. Using this new matrix in Eq. (5.78) leads to another set of ps, so that

$$k_2 = |\Im(p)| \qquad \gamma_2 = \frac{\Re(p)}{k_2} \tag{5.80}$$

Continual updating of the aerodynamic matrix in this way provides an iterative scheme that is convergent for each of the roots, negative γ being a measure of the modal damping. The earliest presentation of this technique was offered by Irwin and Guyett in 1965. For low-order problems, it is straightforward to use a root-finding procedure in which the determinant obtained from setting $k = |\Im(p)|$ is required to vanish.

Hassig applied the p-k method to the configuration in Fig. 5.6. As illustrated by Fig. 5.7 (which is his Fig. 2), the p-k method appears to yield approximately the same result as the p method. This, of course, simply validates the convergence of the scheme. Its greatest advantage is that it can utilize airloads that have been formulated for simple harmonic motion. Another comparison offered by Hassig was between the widely used k method and the p-k method for a horizontal stabilizer/elevator configuration. This example of a strongly coupled system provided the results given

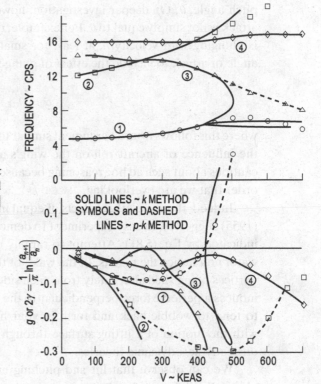

Figure 5.8. Comparison between *p-k* and *k* methods of flutter analysis for a horizontal stabilizer with elevator (from Hassig [1971] Fig. 3, used by permission)

in Fig. 5.8 (which is his Fig. 3). Here again, as in the *k* versus *p* comparison in Fig. 5.6, widely differing conclusions can be drawn regarding the modal coupling. In addition to the easily interpreted frequency and damping plots versus airspeed for strongly coupled systems, a second advantage is offered by the *p-k* method regarding computational effort. The *k* method requires numerous computer runs at constant density to ensure matching of the Mach number with airspeed and altitude. The *p-k* method does not have this requirement.

The accuracy of the *p-k* method depends on the level of damping in any particular mode. It is left as an exercise for readers (see Problem 14) to show that the *p-k* method damping is only a good approximation for the damping in lightly damped modes. Fortunately, these are the modes about which we care the most. Methods presently used in industry are described by Goodman (2001). Currently, most flutter analyses in the aircraft industry are performed using *k* and/or *p-k* methods. Although the *k* method remains popular because of its speed, when accuracy is important and the *p* method is not feasible, industry users seem to favor the *p-k* method, especially those who run the NASTRAN™ aeroelasticity package.

5.5 Unsteady Aerodynamics

In Section 5.2, flutter analysis was conducted using an aerodynamic theory for steady flow. The lift and pitching moment used were functions only of the instantaneous pitch angle, θ. On deeper investigation, however, it is easy to see that the angle of attack is not simply equal to θ. For example, recalling that the airfoil reference point is plunging with velocity $\dot{h}$, at least for small angles, we can justify modifying the angle of attack to include the effect of plunge; viz.

$$\alpha = \theta + \frac{\dot{h}}{U} \tag{5.81}$$

where this follows from an argument similar to the one used in Section 4.2.5 regarding the influence of aircraft roll on the wing's angle of attack. However, we must be cautious about such ad hoc reasoning because there may be other effects of the same order that we are overlooking.

Indeed, there *are* other effects of equal importance that must be included. Fung (1955) suggested an easy experiment to demonstrate that things are not so simple as indicated by Eq. (5.81): Attempt to rapidly move a stick in a straight line through water and notice the results. In the wake of the stick, there is a vortex pattern, with vortices being shed alternately from each side of the stick. This shedding of vortices induces a periodic force perpendicular to the stick's line of motion, causing the stick to tend to wobble back and forth in your hand. A similar phenomenon happens with the motion of a lifting surface through a fluid and must be accounted for in unsteady-aerodynamic theories.

We can observe that lift and pitching moment consist of two parts from two physically different phenomena: noncirculatory and circulatory effects. Circulatory effects are generally more important for aircraft wings. Indeed, in steady flight, it is the circulatory lift that keeps the aircraft aloft. Vortices are an integral part of the process of generation of circulatory lift. Basically, there is a difference in the velocities on the upper and lower surfaces of an airfoil. Such a velocity profile can be represented as a constant velocity flow plus a vortex. In a dynamic situation, the strength of the vortex (i.e., the circulation) is changing with time, as are both the magnitude and direction of the relative wind vector because of airfoil motion. However, the circulatory forces of steady-flow theories do not include the effects of the vortices shed into the wake. Restricting our discussion to two dimensions and potential flow, we recall an implication of the Helmholtz theorem: The total vorticity will always vanish within any closed curve surrounding a particular set of fluid particles. Thus, if a clockwise vorticity develops about the airfoil, a counterclockwise vortex of the same strength must be shed into the flow. As it moves along, this shed vortex changes the flow field by inducing an unsteady flow back onto the airfoil. This behavior is a function of the strength of the shed vortex and its distance away from the airfoil. Thus, accounting for the effect of shed vorticity is, in general, a complex undertaking and would necessitate knowledge of each vortex shed in the flow. However, if we assume that the vortices shed in the flow move with the flow, then we can estimate the effect of these vortices.

Noncirculatory effects, also called apparent mass and inertia effects, are secondary in importance. They are generated when the wing has nonzero acceleration so that it must carry with it some of the surrounding air. That air has finite mass, which leads to inertial forces opposing its acceleration.

In summary, then, unsteady-aerodynamic theories need to account for at least three separate physical phenomena, as follows:

1. Because of the airfoil's unsteady motion relative to the air, the relative wind vector is not fixed in space. This is only partly addressed by corrections such as in Eq. (5.81). The changing direction of the relative wind changes the effective angle of attack and thus changes the lift.
2. As Fung's experiment shows, the airfoil motion disturbs the flow and causes a vortex to be shed at the trailing edge. The downwash from this vortex, in turn, changes the flow that impinges on the airfoil. This unsteady downwash changes the effective angle of attack and thus changes the lift.
3. The motion of the airfoil accelerates air particles near the airfoil surface, thus creating the need to account for the resulting inertial forces (although this "apparent-inertia" effect is less significant than that of the shed vorticity). The apparent-inertia effect does not change the angle of attack but it does, in general, affect both lift and pitching moment.

Additional phenomena that may affect flutter but which are beyond the scope of this text include three-dimensional effects, compressibility, airfoil thickness, flow separation, and stall.

In this section, we present two types of unsteady-aerodynamic theories, both of which are based on potential-flow theory and take into account the effects of shed vorticity, the motion of the airfoil relative to the air, and the apparent-mass effects. The simpler theory is appropriate for classical flutter analysis as well as for the k and p-k methods. The other is a finite-state theory cast in the time domain, appropriate for the eigenvalue analysis involved in the p method as well as for the time-domain analysis required in control design.

5.5.1 Theodorsen's Unsteady Thin-Airfoil Theory

Theodorsen (1934) derived a theory of unsteady aerodynamics for a thin (meaning a flat-plate) airfoil undergoing small, simple harmonic oscillations in incompressible flow. The derivation is based on linear potential-flow theory and is presented in detail along with mathematical subtleties in the textbook by Bisplinghoff, Ashley, and Halfman (1955). The lift contains both circulatory and noncirculatory terms, whereas the pitching moment about the quarter-chord is entirely noncirculatory. According to Theodorsen's theory, the lift and pitching moment are given by

$$
L = 2\pi\rho_\infty U b C(k) \left[\dot{h} + U\theta + b\left(\frac{1}{2} - a\right)\dot{\theta} \right] + \pi\rho_\infty b^2 \left(\ddot{h} + U\dot{\theta} - ba\ddot{\theta} \right)
$$

$$
M_{\frac{1}{4}} = -\pi\rho_\infty b^3 \left[\frac{1}{2}\ddot{h} + U\dot{\theta} + b\left(\frac{1}{8} - \frac{a}{2}\right)\ddot{\theta} \right]
$$

(5.82)

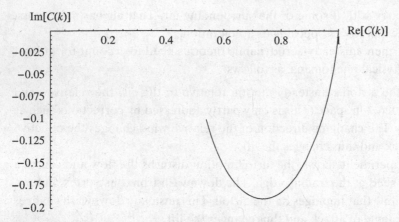

Figure 5.9. Plot of the real and imaginary parts of $C(k)$ for k varying from zero, where $C(k) = 1$, to unity

where the generalized forces are given in Eqs. (5.24). The function $C(k)$ is a complex-valued function of the reduced frequency k, given by

$$C(k) = \frac{H_1^{(2)}(k)}{H_1^{(2)}(k) + i H_0^{(2)}(k)} \tag{5.83}$$

where $H_n^{(2)}(k)$ are Hankel functions of the second kind, which can be expressed in terms of Bessel functions of the first and second kind, respectively, as

$$H_n^{(2)}(k) = J_n(k) - i Y_n(k) \tag{5.84}$$

The function $C(k) = F(k) + i G(k)$ is called Theodorsen's function and is plotted in Figs. 5.9 and 5.10. Note that $C(k)$ is real and equal to unity for the steady case

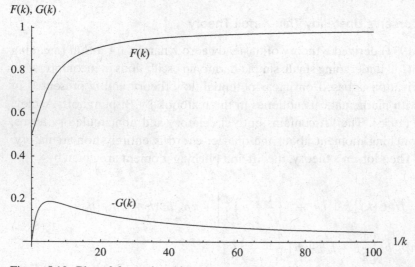

Figure 5.10. Plot of the real and imaginary parts of $C(k)$ versus $1/k$

(i.e., $k = 0$). As k increases, we find that the imaginary part increases in magnitude whereas the real part decreases. As k tends to infinity, $C(k)$ approaches 1/2. However, for practical situations, k rarely exceeds unity. Hence, the plot in Fig. 5.9 only extends to $k = 1$. The large k behavior is shown in Fig. 5.10. When any harmonic function is multiplied by $C(k)$, its magnitude is reduced and a phase lag is introduced. An example of this phenomenon is given herein.

A few things are noteworthy concerning Eqs. (5.82). First, in Theodorsen's theory, the lift-curve slope is equal to 2π. Thus, the first of the two terms in the lift is the circulatory lift without the effect of shed vortices multiplied by $C(k)$. The multiplication by $C(k)$ is a consequence of the theory having considered the effect of shed vorticity. The noncirculatory terms (i.e., the second term in the lift as well as the entire pitching-moment expression) depend on the acceleration and angular acceleration of the airfoil and are mostly apparent-mass/apparent-inertia terms. The circulatory lift is the more significant of the two terms in the lift. Note that the coefficient of $\ddot{h}$ in the lift is the mass per unit length of the air contained in an infinitely long circular cylinder of radius b. This quantity reflects how much air is imparted an acceleration by motion of the airfoil.

For steady flow, the circulatory lift is linear in the angle of attack; however, for unsteady flow, there is no single angle of attack because the flow direction varies along the chordline as the result of the induced flow varying along the chord. However, just so we can discuss the concept for unsteady flow, it is helpful to introduce an effective angle of attack. For simple harmonic motion, it can be inferred from Theodorsen's theory that an effective angle of attack is

$$\alpha = C(k) \left[\theta + \frac{h}{U} + \frac{b}{U} \left(\frac{1}{2} - a \right) \dot{\theta} \right] \tag{5.85}$$

As shown in Section 5.5.2 by comparison with the finite-state aerodynamic model introduced therein, α is the angle of attack measured at the three-quarter chord based on an averaged value of the induced flow over the chord. Recall that in the case of steady-flow aerodynamics of two-dimensional wings, the angle of attack is the pitch angle θ. Here, however, α depends on θ as well as on h, $\dot{\theta}$, and k. Because of these additional terms and because of the behavior of $C(k)$, we expect changes in magnitude and phase between θ and α. These carry over into changes in the magnitude and phase of the lift relative to that of θ. Indeed, the function $C(k)$ is sometimes called the lift-deficiency function because it reduces the magnitude of the unsteady lift relative to the steady lift. It also introduces an important phase shift between the peak values of pitching oscillations and corresponding oscillations in lift.

When we see the dots over h and θ in the lift and pitching-moment expressions, it is tempting to think of them as time-domain equations. However, the presence of $C(k)$ is nonsensical in a time-domain equation. Therefore, Theodorsen's theory with the $C(k)$ present must be recognized as valid only for simple harmonic motion.

Note that an approximation of Theodorsen's theory in which $C(k)$ is set equal to unity is called a "quasi-steady" thin-airfoil theory. Such an approximation has value only for cases in which k is restricted to be very small. For slow harmonic oscillations or slowly varying motion that is not harmonic, the quasi-steady theory may be used in the time domain.

As an example to show the decrease in magnitude and change of phase, consider that the dominant term in the lift is proportional to α. In the time domain, lift is real and so are α and θ. However, when we regard θ as harmonic; viz.

$$\theta = \bar{\theta} \exp(i\omega t) \tag{5.86}$$

then we must realize that to recover the time-domain behavior, we need

$$\theta = \Re[\bar{\theta} \exp(i\omega t)] \tag{5.87}$$

Similarly, we must recover the time-domain behavior of α using the relationship

$$\alpha = \Re[C(k)\bar{\theta} \exp(i\omega t)] \tag{5.88}$$

Now, assuming $\bar{\theta} = 1$ so that in the time domain $\theta = \cos(\omega t)$, we find that

$$\begin{aligned}
\alpha &= \Re[C(k)\bar{\theta} \exp(i\omega t)] \\
&= \Re[C(k) \exp(i\omega t)] \\
&= F(k) \cos(\omega t) - G(k) \sin(\omega t) \\
&= \left[F^2(k) + G^2(k) \right]^{\frac{1}{2}} \cos(\omega t - \phi) \\
&= |C(k)| \cos(\omega t - \phi)
\end{aligned} \tag{5.89}$$

where

$$\tan(\phi) = -\frac{G(k)}{F(k)} \tag{5.90}$$

Because $|C(k)| < 1$ and $\phi(k) > 0$, having the amplitude of θ equal to unity implies that α has an amplitude less than unity; having the peak of θ at $t = 0$ implies α has its peak shifted to $t = \phi/\omega$. For example, when $k = 1/3$, $C(k) = 0.649739 - 0.174712\,i$ so that $|C(k)| = 0.672819$, implying a magnitude reduction of nearly 33%, and $\phi = 15.0506$ degrees.

Theodorsen's theory may be used in classical flutter analysis. There, the reduced frequency of flutter is not known a priori. We can find k at the flutter condition using the method described in Section 5.3. Theodorsen's theory also may be used in the k and p-k methods, as described in Sections 5.4.1 and 5.4.2, respectively.

5.5.2 Finite-State Unsteady Thin-Airfoil Theory of Peters et al.

Although Theodorsen's theory is an excellent choice for classical flutter analysis, there are situations in which an alternative approach is needed. First, we frequently need to calculate the modal damping in subcritical flight conditions. Second, there is a growing interest in the active control of flutter, and design of controllers requires that the system be represented in state-space form. To meet these requirements, we need to represent the actual aerodynamic loads (which are in the frequency

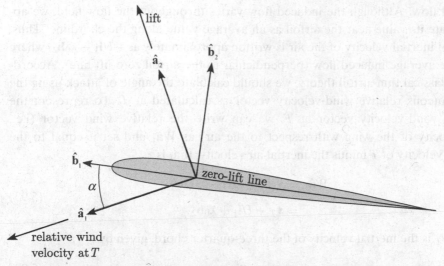

Figure 5.11. Schematic showing geometry of the zero-lift line, relative wind, and lift directions

domain in Theodorsen's theory) in terms of time-domain differential equations. Finite-state theories approximate the actual infinite-state aerodynamic model to within engineering accuracy. One such approach is the finite-state, induced-flow theory for inviscid, incompressible flow of Peters et al. (1995).

Consider a typical section of a rigid, symmetric wing (see Fig. 5.2) and the additional vectorial directions defined in Fig. 5.11. To begin the presentation of this theory, we first relate the three sets of unit vectors, as follows:

1. A set fixed in the inertial frame, $\hat{\mathbf{i}}_1$ and $\hat{\mathbf{i}}_2$, such that the air is flowing at velocity $-U\hat{\mathbf{i}}_1$
2. A set fixed in the wing, $\hat{\mathbf{b}}_1$ and $\hat{\mathbf{b}}_2$, with $\hat{\mathbf{b}}_1$ directed along the zero-lift line toward the leading edge and $\hat{\mathbf{b}}_2$ perpendicular to $\hat{\mathbf{b}}_1$
3. A set $\hat{\mathbf{a}}_1$ and $\hat{\mathbf{a}}_2$ associated with the local relative wind vector at the three-quarter chord, such that $\hat{\mathbf{a}}_1$ is along the relative wind vector and $\hat{\mathbf{a}}_2$ is perpendicular to it, in the assumed direction of the lift

The relationships among these unit vectors can be stated simply as

$$\begin{Bmatrix} \hat{\mathbf{b}}_1 \\ \hat{\mathbf{b}}_2 \end{Bmatrix} = \begin{bmatrix} \cos(\theta) & \sin(\theta) \\ -\sin(\theta) & \cos(\theta) \end{bmatrix} \begin{Bmatrix} \hat{\mathbf{i}}_1 \\ \hat{\mathbf{i}}_2 \end{Bmatrix} \tag{5.91}$$

and

$$\begin{Bmatrix} \hat{\mathbf{a}}_1 \\ \hat{\mathbf{a}}_2 \end{Bmatrix} = \begin{bmatrix} \cos(\alpha) & -\sin(\alpha) \\ \sin(\alpha) & \cos(\alpha) \end{bmatrix} \begin{Bmatrix} \hat{\mathbf{b}}_1 \\ \hat{\mathbf{b}}_2 \end{Bmatrix} \tag{5.92}$$

and $\hat{\mathbf{i}}_3 = \hat{\mathbf{a}}_3 = \hat{\mathbf{b}}_3 = \hat{\mathbf{b}}_1 \times \hat{\mathbf{b}}_2$.

Induced-flow theories approximate the effects of shed vortices based on changes they cause in the flow field near the airfoil. Thus, the velocity field near the airfoil consists of the freestream velocity plus an additional component to account for the induced flow. Although the induced flow varies throughout the flow field, we approximate its value near the airfoil as an average value along the chordline. Thus, the local inertial velocity of the air is written approximately as $-U\hat{\mathbf{i}}_1 - \lambda_0\hat{\mathbf{b}}_2$, where λ_0 is the average induced flow (perpendicular to the airfoil zero-lift line). According to classical thin-airfoil theory, we should calculate the angle of attack using the instantaneous relative wind-velocity vector as calculated at T. To represent the relative wind-velocity vector at T, we can write the relative wind vector (i.e., the velocity of the wing with respect to the air) as $W\hat{\mathbf{a}}_1$ and set it equal to the inertial velocity of T minus the inertial air velocity; that is

$$
\begin{aligned}
W\hat{\mathbf{a}}_1 &= \mathbf{v}_T - (-U\hat{\mathbf{i}}_1 - \lambda_0\hat{\mathbf{b}}_2) \\
&= \mathbf{v}_T + U\hat{\mathbf{i}}_1 + \lambda_0\hat{\mathbf{b}}_2
\end{aligned}
\tag{5.93}
$$

where $\mathbf{v}_T$ is the inertial velocity of the three-quarter chord, given by

$$
\mathbf{v}_T = \mathbf{v}_P + \dot{\theta}\hat{\mathbf{b}}_3 \times \mathbf{r}_{PT}
\tag{5.94}
$$

and $\mathbf{r}_{PT}$ is the position vector from P to T. Fig. 5.2 shows that

$$
\mathbf{r}_{PT} = \left[\frac{b}{2} + (1+a)b - 2b\right]\hat{\mathbf{b}}_1 = b\left(a - \frac{1}{2}\right)\hat{\mathbf{b}}_1
\tag{5.95}
$$

Thus,

$$
\mathbf{v}_T = \mathbf{v}_P + \dot{\theta}\hat{\mathbf{b}}_3 \times b\left(a - \frac{1}{2}\right)\hat{\mathbf{b}}_1
\tag{5.96}
$$

The inertial velocity of the reference point P is

$$
\mathbf{v}_P = -\dot{h}\hat{\mathbf{i}}_2
\tag{5.97}
$$

whereas $\dot{\theta}\hat{\mathbf{b}}_3$ is the inertial angular velocity of the wing. Carrying out the cross product in Eq. (5.96), we obtain

$$
\mathbf{v}_T = -\dot{h}\hat{\mathbf{i}}_2 + b\dot{\theta}\left(a - \frac{1}{2}\right)\hat{\mathbf{b}}_2
\tag{5.98}
$$

so that the relative wind can be written as

$$
W\hat{\mathbf{a}}_1 = U\hat{\mathbf{i}}_1 - \dot{h}\hat{\mathbf{i}}_2 + \left[b\dot{\theta}\left(a - \frac{1}{2}\right) + \lambda_0\right]\hat{\mathbf{b}}_2
\tag{5.99}
$$

Alternatively, we may write the relative wind in terms of its components along $\hat{\mathbf{b}}_1$ and $\hat{\mathbf{b}}_2$; that is

$$
W\hat{\mathbf{a}}_1 = W\cos(\alpha)\hat{\mathbf{b}}_1 - W\sin(\alpha)\hat{\mathbf{b}}_2
\tag{5.100}
$$

where α is given by (see Fig. 5.11)

$$
\tan(\alpha) = -\frac{\hat{\mathbf{a}}_1 \cdot \hat{\mathbf{b}}_2}{\hat{\mathbf{a}}_1 \cdot \hat{\mathbf{b}}_1}
\tag{5.101}
$$

Using Eq. (5.99), we find that

$$W\hat{\mathbf{a}}_1 \cdot \hat{\mathbf{b}}_1 = U\cos(\theta) - h\sin(\theta)$$

$$W\hat{\mathbf{a}}_1 \cdot \hat{\mathbf{b}}_2 = -U\sin(\theta) - h\cos(\theta) + b\left(a - \frac{1}{2}\right)\dot{\theta} + \lambda_0 \tag{5.102}$$

Assuming small angles, we now may show that

$$\alpha = \theta + \frac{h}{U} + \frac{b}{U}\left(\frac{1}{2} - a\right)\dot{\theta} - \frac{\lambda_0}{U} \tag{5.103}$$

$$W = U + \text{ higher-order terms}$$

According to this derivation, α is an effective angle of attack based on the relative wind vector at the three-quarter chord, which, in turn, is based on the average value of the induced flow λ_0 over the wing chordline. Note that α is *not* equal to the pitch angle θ. Because of the motion of the wing and the induced flow field, the relative wind direction is not fixed in inertial space. Therefore, the effective angle of attack depends on the pitch rate, the plunge velocity, and the induced flow. Moreover, the lift is assumed to be perpendicular to the relative wind vector. This assumption is adequate for the calculation of lift and pitching moment, which are both first-order in the motion variables. However, sufficiently rapid plunge motion (e.g., as in the flapping wings of an insect) can result in a value of α that is not small, and we would need to make "small but finite" angle assumptions to calculate the drag (or propulsive force equal to negative drag) that could be encountered in such situations.

The total lift and moment expressions including the noncirculatory forces are

$$L = \pi\rho_\infty b^2\left(\ddot{h} + U\dot{\theta} - ba\ddot{\theta}\right) + 2\pi\rho_\infty Ub\left[h + U\theta + b\left(\frac{1}{2} - a\right)\dot{\theta} - \lambda_0\right] \tag{5.104}$$

$$M_{\frac{1}{4}} = -\pi\rho_\infty b^3\left[\frac{1}{2}\ddot{h} + U\dot{\theta} + b\left(\frac{1}{8} - \frac{a}{2}\right)\ddot{\theta}\right]$$

Note the similarity between Eqs. (5.104) and (5.82). In particular, by studying the circulatory lift in both lift equations, we then can see the basis for identifying α, calculated as in the first of Eqs. (5.103) with the expression in Eq. (5.85).

The lift and pitching moment then are used to form the generalized forces from Eqs. (5.24) and, in turn, are used in the structural equations in Eqs. (5.26). Even so, these two equations are incomplete, having more than two unknowns. The induced-flow velocity λ_0 must be expressed in terms of the airfoil motion. The induced-flow theory of Peters et al. does that, representing the average induced-flow velocity λ_0 in terms of N induced-flow states $\lambda_1, \lambda_2, \ldots, \lambda_N$ as

$$\lambda_0 \approx \frac{1}{2}\sum_{n=1}^{N} b_n\lambda_n \tag{5.105}$$

where the b_n are found by the least-squares method. The induced-flow dynamics then are derived from the assumption that the shed vortices stay in the plane of the airfoil and travel downstream with the same velocity as the flow. Introducing a column matrix $\{\lambda\}$ containing the values of λ_n, we can write the set of N first-order ordinary differential equations governing $\{\lambda\}$ as

$$[A]\{\dot{\lambda}\} + \frac{U}{b}\{\lambda\} = \{c\}\left[\ddot{h} + U\dot{\theta} + b\left(\frac{1}{2} - a\right)\ddot{\theta}\right] \tag{5.106}$$

where the matrices $[A]$ and $\{c\}$ can be derived for a user-defined number of induced-flow states. The expressions of the matrices used here are given for N finite states as

$$[A] = [D] + \{d\}\{b\}^T + \{c\}\{d\}^T + \frac{1}{2}\{c\}\{b\}^T \tag{5.107}$$

where

$$
\begin{aligned}
D_{nm} &= \tfrac{1}{2n} & n &= m+1 \\
&= -\tfrac{1}{2n} & n &= m-1 \\
&= 0 & n &\neq m \pm 1
\end{aligned} \tag{5.108}
$$

$$
\begin{aligned}
b_n &= (-1)^{n-1}\frac{(N+n-1)!}{(N-n-1)!}\frac{1}{(n!)^2} & n &\neq N \\
&= (-1)^{n-1} & n &= N
\end{aligned} \tag{5.109}
$$

$$
\begin{aligned}
d_n &= \tfrac{1}{2} & n &= 1 \\
&= 0 & n &\neq 1
\end{aligned} \tag{5.110}
$$

and

$$c_n = \frac{2}{n} \tag{5.111}$$

The resulting aeroelastic model is in the time domain, in contrast to classical flutter analysis, which is in the frequency domain (see Section 5.3). Thus, it can be used for flutter analysis by the p method, as well as in the design of control systems to alleviate flutter.

Results using the finite-state, induced-flow model (i.e., Eqs. 5.106 and 5.26 with generalized forces given by Eqs. 5.24 with lift and pitching moment given by Eqs. 5.104) for the problem analyzed previously in Section 5.2 (recall that $a = -1/5$, $e = -1/10$, $\mu = 20$, $r^2 = 6/25$, and $\sigma = 2/5$) are given here. These results are based on use of $N = 6$ induced-flow states.[2] The frequency and damping results are shown in Figs. 5.12 and 5.13, respectively. As before, a frequency coalescence is observed near the instability, but the flutter condition is marked by the crossing of the real part of one of the roots into positive territory. The flutter speed obtained is $V_F = U_F/(b\omega_\theta) = 2.165$, and the flutter frequency is $\Omega_F/\omega_\theta = 0.6545$. Although this value of the flutter speed is close to that observed previously using the simpler theory, the unsteady-aerodynamics theory produces complex roots for all $V \neq 0$ so that there is modal damping in all of the modes below the flutter speed. The equations contain damping terms proportional to the velocity that account for the initial increase in

---

[2] It should be noted that a larger number of induced-flow states is not necessary. The use of too many may degrade the accuracy of the model because of ill-conditioning.

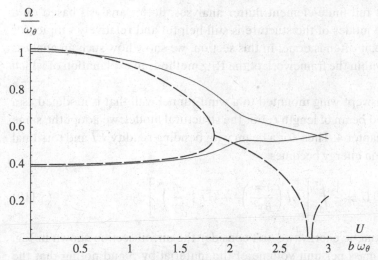

Figure 5.12. Plot of the modal frequency versus $U/(b\omega_\theta)$ for $a = -1/5$, $e = -1/10$, $\mu = 20$, $r^2 = 6/25$, and $\sigma = 2/5$; solid lines: p method, aerodynamics of Peters et al.; dashed lines: steady-flow aerodynamics

damping. At higher velocities, however, the destabilizing circulatory term (i.e., the nonsymmetric term in the stiffness matrix that also is present in the steady-flow theory) overcomes the damping caused by the unsteady terms, resulting in flutter.

It is left to readers as an exercise to show the equivalence of the theories of Peters et al. and Theodorsen (see Problem 16).

5.6 Flutter Prediction via Assumed Modes

As previously noted, in industry it is now typical to use the finite element method as a means to realistically represent aircraft structural dynamics. Although it is certainly

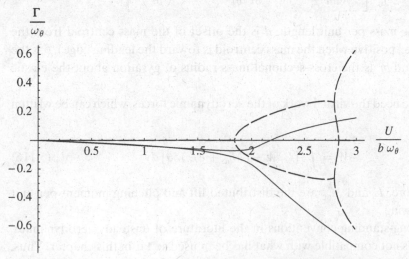

Figure 5.13. Plot of the modal damping versus $U/(b\omega_\theta)$ for $a = -1/5$, $e = -1/10$, $\mu = 20$, $r^2 = 6/25$, and $\sigma = 2/5$; solid lines: p method, aerodynamics of Peters et al.; dashed lines: steady-flow aerodynamics

possible to conduct full finite-element flutter analyses, flutter analysis based on a truncated set of the modes of the stucture is still helpful and relatively simple; for those reasons alone, it often is done. In this section, we show how such an analysis can be performed within the framework of the Ritz method, an explanation of which is in Section 3.5.

Consider an unswept wing mounted to a wind-tunnel wall that is modeled as a uniform cantilevered beam of length ℓ. For the structural model, we adopt the same notation used in Chapter 4. Thus, for a beam with bending rigidity $\overline{EI}$ and torsional rigidity $\overline{GJ}$, the strain energy becomes

$$U = \frac{1}{2} \int_0^\ell \left[\overline{EI} \left(\frac{\partial^2 w}{\partial y^2} \right)^2 + \overline{GJ} \left(\frac{\partial \theta}{\partial y} \right)^2 \right] dy \tag{5.112}$$

To obtain the kinetic energy, we first consider the airfoil section shown in Fig. 4.12. Denoting the mass per unit volume of the material by ρ and noting that the velocity of a typical point within the cross-sectional plane is

$$\mathbf{v} = z \frac{\partial \theta}{\partial t} \hat{\mathbf{i}} + \left(\frac{\partial w}{\partial t} - x \frac{\partial \theta}{\partial t} \right) \hat{\mathbf{k}} \tag{5.113}$$

where $\hat{\mathbf{i}}$ and $\hat{\mathbf{k}}$ are unit vectors in the x and z directions, respectively, we can write the kinetic energy as

$$K = \frac{1}{2} \int_0^\ell \iint_A \rho \left[\left(\frac{\partial w}{\partial t} - x \frac{\partial \theta}{\partial t} \right)^2 + z^2 \left(\frac{\partial \theta}{\partial t} \right)^2 \right] dx\, dz\, dy \tag{5.114}$$

Straightforward evaluation of the cross-sectional integrals yields

$$K = \frac{1}{2} \int_0^\ell \left[m \left(\frac{\partial w}{\partial t} \right)^2 + 2md \frac{\partial w}{\partial t} \frac{\partial \theta}{\partial t} + mb^2 r^2 \left(\frac{\partial \theta}{\partial t} \right)^2 \right] dy \tag{5.115}$$

where m is the mass per unit length, d is the offset of the mass centroid from the elastic axis (i.e., positive when the mass centroid is toward the leading edge), b is the semi-chord, and br is the cross-sectional mass radius of gyration about the elastic axis.

Finally, we need the virtual work of the aerodynamic forces, which can be written as

$$\overline{\delta W} = \int_0^\ell \left[L' \delta w + (M'_{ac} + eL') \delta\theta \right] dy \tag{5.116}$$

where, as before, L' and M'_{ac} are the distributed lift and pitching moment per unit length of the wing.

Due to long-standing conventions in the literature of unsteady aerodynamics, this notation is not compatible with what has been used so far in this chapter. Thus, we rewrite these three expressions (i.e., strain energy, kinetic energy, and virtual work) in terms of the notation of this chapter. In particular, we can show that the following replacements can be made for the notation used in Fig. 4.12:

$$d \rightarrow -bx_\theta$$

$$e \rightarrow \left(\frac{1}{2} + a\right) b$$

$$L' \rightarrow L'$$

$$M'_{\text{ac}} \rightarrow M'_{\frac{1}{4}}$$

Thus, the strain energy is unchanged from before. The kinetic energy becomes

$$K = \frac{1}{2} \int_0^\ell \left[m \left(\frac{\partial w}{\partial t}\right)^2 - 2mbx_\theta \frac{\partial w}{\partial t} \frac{\partial \theta}{\partial t} + mb^2 r^2 \left(\frac{\partial \theta}{\partial t}\right)^2 \right] dy \qquad (5.117)$$

and the virtual work becomes

$$\overline{\delta W} = \int_0^\ell \left\{ L' \delta w + \left[M'_{\frac{1}{4}} + \left(\frac{1}{2} + a\right) b L' \right] \delta \theta \right\} dy \qquad (5.118)$$

A reasonable choice for the assumed modes is the set of uncoupled cantilevered-beam, free-vibration modes for bending and torsion, such that

$$w(y, t) = \sum_{i=1}^{N_w} \eta_i(t) \Psi_i(y)$$

$$\theta(y, t) = \sum_{i=1}^{N_\theta} \phi_i(t) \Theta_i(y) \qquad (5.119)$$

where N_w and N_θ are the numbers of modes used to represent bending and torsion, respectively; η_i and ϕ_i are the generalized coordinates associated with bending and torsion, respectively; and Ψ_i and Θ_i are the bending and torsion mode shapes, respectively. Here, Θ_i is given by

$$\Theta_i = \sqrt{2} \sin(\gamma_i y) \qquad (5.120)$$

where

$$\gamma_i = \frac{\pi \left(i - \frac{1}{2}\right)}{\ell} \qquad (5.121)$$

and, according to Eq. (3.258), Ψ_i is given as

$$\Psi_i = \cosh(\alpha_i y) - \cos(\alpha_i y) - \beta_i [\sinh(\alpha_i y) - \sin(\alpha_i y)] \qquad (5.122)$$

with α_i and β_i as given in Table 3.1.

The next step in the application of the Ritz method is to discretize spatially the expressions for strain energy, kinetic energy, and virtual work. Because of the orthogonality of both the bending and torsion modes, the strain energy simplifies to

$$U = \frac{1}{2} \left[\frac{EI}{\ell^3} \sum_{i=1}^{N_w} (\alpha_i \ell)^4 \eta_i^2 + \frac{GJ}{\ell} \sum_{i=1}^{N_\theta} (\gamma_i \ell)^2 \phi_i^2 \right] \qquad (5.123)$$

Similarly, the kinetic energy is simplified considerably because of the ortho-gonality of both the bending and torsion modes and can be written as

$$K = \frac{m\ell}{2} \left(\sum_{i=1}^{N_w} \dot{\eta}_i^2 + b^2 r^2 \sum_{i=1}^{N_\theta} \dot{\phi}_i^2 - 2bx_\theta \sum_{i=1}^{N_\theta} \sum_{j=1}^{N_w} A_{ij} \dot{\phi}_i \dot{\eta}_j \right) \qquad (5.124)$$

where

$$A_{ij} = \frac{1}{\ell} \int_0^\ell \Theta_i \Psi_j \, dy \qquad i = 1, 2, \ldots, N_\theta \qquad j = 1, 2, \ldots, N_w \qquad (5.125)$$

Inertial coupling between bending and torsion motion is reflected by the term in-volving A_{ij}, which is a fully populated matrix because the bending and torsion modes are not orthogonal to one another.

The virtual-work expression

$$\overline{\delta W} = \sum_{i=1}^{N_w} \Xi_{w_i} \delta \eta_i + \sum_{i=1}^{N_\theta} \Xi_{\theta_i} \delta \phi_i \qquad (5.126)$$

can be used to identify the generalized forces. Thus

$$\Xi_{w_i} = \int_0^\ell \Psi_i L' \, dy$$

$$\Xi_{\theta_i} = \int_0^\ell \Theta_i \left[M'_{\frac{1}{4}} + \left(\frac{1}{2} + a \right) b L' \right] dy \qquad (5.127)$$

where expressions for L' and $M'_{\frac{1}{4}}$ can be found by taking expressions for L and $M_{\frac{1}{4}}$ in Eqs. (5.82) or (5.104) and replacing h with $-w$ and dots with partial derivatives with respect to time. This we carry out for illustrative purposes using Theodorsen's theory, for which

$$L' = 2\pi \rho_\infty U b C(k) \left[U\theta - \frac{\partial w}{\partial t} + b \left(\frac{1}{2} - a \right) \frac{\partial \theta}{\partial t} \right] + \pi \rho_\infty b^2 \left(U \frac{\partial \theta}{\partial t} - \frac{\partial^2 w}{\partial t^2} - ba \frac{\partial^2 \theta}{\partial t^2} \right)$$

$$M'_{\frac{1}{4}} = -\pi \rho_\infty b^3 \left[U \frac{\partial \theta}{\partial t} - \frac{1}{2} \frac{\partial^2 w}{\partial t^2} + b \left(\frac{1}{8} - \frac{a}{2} \right) \frac{\partial^2 \theta}{\partial t^2} \right] \qquad (5.128)$$

Substituting Eqs. (5.119) into Eqs. (5.128), we obtain expressions for the generalized forces that can be put easily into matrix form:

$$\begin{Bmatrix} \Xi_w \\ \Xi_\theta \end{Bmatrix} = -\pi \rho_\infty b^2 \ell \begin{bmatrix} [\Delta] & ba[A]^T \\ ba[A] & b^2 \left(a^2 + \frac{1}{8} \right) [\Delta] \end{bmatrix} \begin{Bmatrix} \ddot{\eta} \\ \ddot{\phi} \end{Bmatrix}$$

$$-\pi \rho_\infty b U \ell \begin{bmatrix} 2C(k)[\Delta] & -b \left[1 + 2 \left(\frac{1}{2} - a \right) C(k) \right] [A]^T \\ 2b \left(\frac{1}{2} + a \right) C(k)[A] & b^2 \left(\frac{1}{2} - a \right) \left[1 - 2 \left(\frac{1}{2} + a \right) C(k) \right] [\Delta] \end{bmatrix} \begin{Bmatrix} \dot{\eta} \\ \dot{\phi} \end{Bmatrix}$$

$$-\pi \rho_\infty b U^2 \ell \begin{bmatrix} [0] & -2C(k)[A]^T \\ [0] & -b(1 + 2a)C(k)[\Delta] \end{bmatrix} \begin{Bmatrix} \eta \\ \phi \end{Bmatrix} \qquad (5.129)$$

where $[\Delta]$ denotes an identity matrix and $[0]$ denotes a matrix of zeros. Because of limitations inherent in the derivation of Theodorsen's theory, this expression for the generalized forces is valid only for simple harmonic motion. Note that the rectangular submatrices in this equation are also referred to as aerodynamic-influence coefficients (AICs).

All that now remains in the application of the Ritz method is to invoke Lagrange's equations to obtain the generalized equations of motion, which can be written in matrix form as

$$
m\ell \begin{bmatrix} [\Delta] & -bx_\theta[A]^T \\ -bx_\theta[A] & b^2 r^2[\Delta] \end{bmatrix} \begin{Bmatrix} \ddot{\eta} \\ \ddot{\phi} \end{Bmatrix} + \begin{bmatrix} \frac{EI}{\ell^3}\lceil B. \rfloor & [0] \\ [0] & \frac{GJ}{\ell}\lceil T. \rfloor \end{bmatrix} \begin{Bmatrix} \eta \\ \phi \end{Bmatrix} = \begin{Bmatrix} \Xi_w \\ \Xi_\theta \end{Bmatrix} \tag{5.130}
$$

where elements of the diagonal matrices $\lceil B. \rfloor$ and $\lceil T. \rfloor$ are given by

$$
\begin{aligned}
B_{ii} &= (\alpha_i \ell)^4 \\
T_{ii} &= (\gamma_i \ell)^2
\end{aligned} \tag{5.131}
$$

The appearance of diagonal matrices $\lceil B. \rfloor$ and $\lceil T. \rfloor$ in the stiffness matrix and the appearances of Δ in the mass matrix and generalized forces are caused by the orthogonality of the chosen basis functions Θ_i and Ψ_i. Such a choice is not necessary but it simplifies the discretized equations.

Following the methodology of classical flutter analysis in Section 5.3, we set

$$
\begin{aligned}
\eta(t) &= \bar{\eta} \exp(i\omega t) \\
\phi(t) &= \bar{\phi} \exp(i\omega t)
\end{aligned} \tag{5.132}
$$

where ω is the frequency of the simple harmonic motion. This leads to a flutter determinant that can be solved by following steps similar to those outlined in Section 5.3, the only difference being that there are now more degrees of freedom if either N_w or N_θ exceeds unity.

Let us consider the case in which $N_w = N_\theta = 1$. If we introduce dimensionless constants similar to those in Section 5.3, the equations of motion can be put in the form of Eqs. (5.62); that is

$$
\begin{aligned}
\left\{ \mu \left[1 - \left(\frac{\omega_w}{\omega} \right)^2 \right] + \ell_w \right\} \frac{\bar{\eta}_1}{b} + (-\mu x_\theta + \ell_\theta) A_{11} \bar{\phi}_1 &= 0 \\
(-\mu x_\theta + m_w) A_{11} \frac{\bar{\eta}_1}{b} + \left\{ \mu r^2 \left[1 - \left(\frac{\omega_\theta}{\omega} \right)^2 \right] + m_\theta \right\} \bar{\phi}_1 &= 0
\end{aligned} \tag{5.133}
$$

Here, ℓ_w, ℓ_θ, m_w, and m_θ are defined in a manner similar to the quantities on the right-hand side of Eqs. (5.39) with the loads from Theodorsen's theory

$$\ell_w = 1 - \frac{2iC(k)}{k}$$

$$\ell_\theta = a + \frac{i}{k}\left[1 + 2\left(\frac{1}{2} - a\right)C(k)\right] + \frac{2C(k)}{k^2}$$

$$m_w = a - \frac{2i\left(\frac{1}{2} + a\right)C(k)}{k} \tag{5.134}$$

$$m_\theta = a^2 + \frac{1}{8} - \frac{\left(\frac{1}{2} - a\right)\left[1 - 2\left(\frac{1}{2} + a\right)C(k)\right]i}{k} + \frac{2\left(\frac{1}{2} + a\right)C(k)}{k^2}$$

and the fundamental bending and torsion frequencies are

$$\omega_w = (\alpha_1\ell)^2\sqrt{\frac{EI}{m\ell^4}}$$

$$\omega_\theta = \frac{\pi}{2}\sqrt{\frac{GJ}{mb^2r^2\ell^2}} \tag{5.135}$$

Finally, the constant $A_{11} = 0.958641$. It is clear that these equations are in the same form as those solved previously for the typical section and that the influence of wing flexibility for this simplest two-mode case enters only in a minor way—namely, to adjust the coupling terms by a factor of less than 5%.

The main purpose of this example is to demonstrate how the tools already presented can be used to conduct a flutter analysis of a flexible wing. Addition of higher modes certainly can affect the results, as can such things as spanwise variations in the mass and stiffness properties and concentrated masses and inertias along the wing. Incorporation of these additional features into the analysis would make the analysis more suitable for realistic flutter calculations.

However, to fully capture the realism afforded by these and other important considerations—such as aircraft with delta-wing configurations or very-low-aspect-ratio-wings—a full finite-element analysis is necessary. Even in such cases, it is typical that flutter analyses based on assumed modes give analysts a reasonably good idea of the mechanisms of instability. Moreover, the full finite element method can be used to obtain a realistic set of modes that, in turn, could be used in a Ritz-method analysis instead of those used herein. This way, considerable realism can be incorporated into the model without necessitating the model to be of large order. Present-day industry practice uses both full finite-element models as well as assumed modes derived from a full finite-element model.

As for the unsteady aerodynamics, in industry, the AIC matrices usually are computed using panel codes based on unsteady potential flow, such as the doublet-lattice method. The geometry of the panels, in general, is quite different from that of the structural-finite elements. This gives rise to the need for transferring both motion and loads between these two models. One approach for transferal uses a spline matrix that

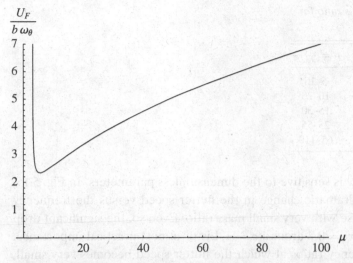

Figure 5.14. Plot of dimensionless flutter speed versus mass ratio for the case $\sigma = 1/\sqrt{10}$, $r = 1/2$, $x_\theta = 0$, and $a = -3/10$

interpolates the displacements at structural-finite-element grid points to those at the panels of the aerodynamics code and transfers loads on the panels to the nodes of the finite elements. Methodology has been developed that fosters straightforward coupling of structural and aerodynamic codes despite disparities in their meshes (Smith, Cesnik, and Hodges, 1995). Similar procedures also can be used to couple finite-element codes with more sophisticated computational fluid dynamics (CFD) codes.

5.7 Flutter Boundary Characteristics

The preceding sections describe procedures for the determination of the flutter boundary in terms of altitude, speed, and Mach number. For a standard atmosphere, any two of these conditions are sufficient to describe the flight condition. The final flutter boundary is presented frequently in terms of a dimensionless flutter speed as $U_F/(b\omega_\theta)$. The parameter $U/(b\omega_\theta)$ sometimes is referred to as the reduced velocity, although the reciprocal of the reduced frequency $U/(b\omega)$ is also sometimes so designated. A useful presentation of this reduced flutter speed as a function of the mass ratio, $\mu = m/(\pi\rho_\infty b^2)$, is illustrated in Fig. 5.14. It is immediately apparent that the flutter speed increases in a nearly linear manner with increasing mass ratio. This result can be interpreted in either of two ways. For a given configuration, variations in μ would correspond to changes in atmospheric density and, therefore, altitude. In such a case, the mass ratio increases with increasing altitude. This implies that any flight vehicle is more susceptible to aeroelastic flutter at low rather than higher altitudes.

A second interpretation of the mass ratio is related to its numerical value for any fixed altitude. The value of μ depends on the type of flight vehicle, as reflected by the mass per unit span of the lifting surface, m. Table 5.2 lists vehicle configurations and typical mass-ratio values for atmospheric densities between sea level and 10,000 feet.

Table 5.2. *Variation of mass ratio for typical vehicle types*

Vehicle type	$\mu = \frac{m}{\pi \rho_\infty b^2}$
Gliders and Ultralights	5–15
General Aviation	10–20
Commercial Transports	15–30
Attack Aircraft	25–55
Helicopter Blades	65–110

The flutter boundary is sensitive to the dimensionless parameters. In Fig. 5.15, for example, we see a dramatic change in the flutter speed versus the frequency ratio $\sigma = \omega_h/\omega_\theta$ for a case with very small mass ratio. Even so, the significant drop in the flutter speed for $x_\theta = 0.2$ around $\sigma = 1.4$ is of utmost practical importance. There are certain frequency ratios at which the flutter speed becomes very small, depending on the values of the other parameters. This dip is observed in the plot of flutter speed versus frequency ratio for the wings of most high-performance aircraft, which have relatively large mass ratios and positive static unbalances. The chordwise offsets also have a strong influence on the flutter speed, as shown in Fig. 5.16. Indeed, a small change in the mass-center location can lead to a large increase in the flutter speed. The mass-center location, e, cannot be changed without simultaneously changing the dimensionless radius of gyration, r; however, the relative change in the flutter speed for a small percentage change in the former is more than for a similar percentage change in the latter. These facts led to a concept of mass-balancing wings to alleviate flutter, similar to the way that control surfaces are mass-balanced. If the center of mass is moved forward of the reference point, the flutter speed is generally

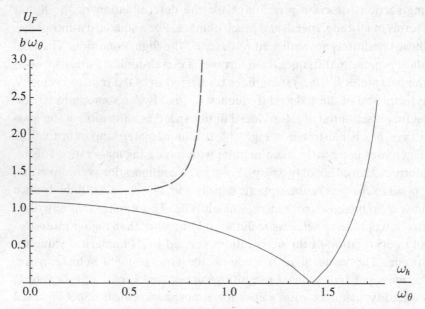

Figure 5.15. Plot of dimensionless flutter speed versus frequency ratio for the case $\mu = 3$, $r = 1/2$, and $a = -1/5$, where the solid line is for $x_\theta = 0.2$ and the dashed line is for $x_\theta = 0.1$

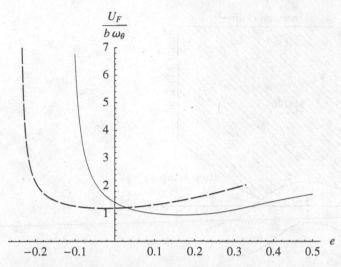

Figure 5.16. Plot of dimensionless flutter speed versus e for the case $\mu = 10$, $\sigma = 1/\sqrt{2}$, and $r = 1/2$; the solid line is for $a = 0$ and the dashed line is for $a = 0.2$

relatively high. Unfortunately, this is not easily accomplished; however, a large change is not usually needed to ensure safety. Note that care must be exercised in examining changes in other parameters caused by such changes in the mass distribution. For example, the torsional frequency may be altered significantly in the process of changing the radius of gyration. Finally, we note that the flutter frequency for bending-torsion flutter is somewhere between ω_h and ω_θ, where normally $\sigma < 1$; however, situations arise in which the flutter frequency may exceed ω_θ.

It is important to note that there are some combinations of the chordwise offset parameters e and a for which the current simplified theories indicate that flutter is not possible. The classic textbook by Bisplinghoff, Ashley, and Halfman (1955) classified the effects of the chordwise offsets e and a in terms of small and large σ. For small σ, they noted that flutter can happen only when the mass center is behind the quarter-chord (i.e., when $e > -1/2$); thus, it cannot happen when $e \le -1/2$. For large σ, flutter can happen only when the elastic axis is in front of the quarter-chord (i.e., when $a < -1/2$); thus, it cannot happen when $a \ge -1/2$. Moreover, for the typical-section model in combination with the aerodynamic models presented herein, flutter does not appear to happen for any combination of σ and r when the mass centroid, elastic axis, and aerodynamic center all coincide (i.e., when $e = a = -1/2$). Even if this prediction of the analysis is correct, practically speaking, it is difficult to achieve coincidence of these points in wing design. Remember, however, that all of these statements are made with respect to simplified models. We need to analyze real wings in a design setting using powerful tools, such as NASTRAN™ or ASTROS™. Indeed, bending-torsion flutter is a complicated phenomenon and it seems to defy all of our attempts at generalization. Additional discussion of these phenomena, along with a large body of solution plots, is found in Bisplinghoff, Ashley, and Halfman (1955).

The final flutter boundary can be presented in numerous ways for any given flight vehicle. The manner in which it is illustrated depends on the engineering purpose that it is intended to serve. One possible presentation of the flutter boundary is to

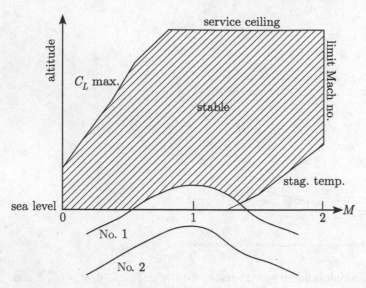

Figure 5.17. Flight envelope for typical Mach 2 fighter

The final flutter boundary can be presented in numerous ways for any given flight vehicle. The manner in which it is illustrated depends on the engineering purpose that it is intended to serve. One possible presentation of the flutter boundary is to superpose it on the vehicle's flight envelope. A typical flight envelope for a Mach 2 attack aircraft is illustrated in Fig. 5.17 with two flutter boundaries indicated by the curves marked No. 1 and No. 2. The shaded region above the flutter boundaries, being at higher altitudes, corresponds to stable flight conditions; below the boundaries, flutter will be experienced. Flutter boundary No. 1 indicates that for a portion of the intended flight envelope, the vehicle will experience flutter. Note that these conditions of instability correspond to a flight Mach number near unity (i.e., transonic flow) and high dynamic pressure. This observation can be generalized by stating that a flight vehicle is more susceptible to aeroelastic flutter for conditions of (1) lower altitude, (2) transonic flow, and (3) higher dynamic pressure.

If it is determined that the vehicle will experience flutter in any portion of its intended flight envelope, it is necessary to make appropriate design changes to eliminate the instability for such conditions. These changes may involve alteration of the inertial, elastic, or aerodynamic properties of the configuration; often, small variations in all three provide the best compromise. Flutter boundary No. 2 is indicative of a flutter-safe vehicle. Note that at the minimum altitude-transonic condition, there appears to be a safety margin with respect to flutter instability. All flight-vehicle specifications require such a safety factor, which is generally called the "flutter margin." Specifications on U.S. military and commercial transport aircraft require that the margin be 15% over the limit equivalent airspeed. In other words, the minimum flutter speed at sea level should not be less than 1.15 times the airspeed for the maximum expected dynamic pressure as evaluated at sea level. For general aviation aircraft the margin is 20%. There may be other values in foreign military requirements. These values are for a normal configuration.

5.8 Structural Dynamics, Aeroelasticity, and Certification[3]

So, with all of this background on theoretical methods, what are some of the ways aeroelasticity and structural dynamics analyses are actually used? We must recall that for every aircraft, there may be dozens to several hundred combinations of fuel and payloads that must be verified as stable within the aircraft's flight envelope before clearance for flight is given. The use of computational results is crucial because we cannot possibly test every combination of fuel and hardware mounted on the fuselage and wings (e.g., stores, armaments, fuel tanks). Computational results then become our main work tool for every go/no-go decision made in flight-testing and ultimate airplane certification for flight.

To proceed with this monumental set of tasks, we first need to identify the most critical combinations (i.e., those with the lowest flutter speeds). If possible, these should be compared with previous experience in terms of computation and flight-testing. Once the most critical configurations are identified, we set them aside for special wind-tunnel and flight tests. In particular, we need to ascertain the flutter mode's shape, frequency ω_F, speed U_F, and severity $g' = dg/dU$, all evaluated at the flutter speed (i.e., where g vanishes). Identification of all four items allows us to distinguish between various cases with comparable flutter speeds and, together with previous experience, to decide about further needed ground and flight tests to verify computations and flight clearance.

Analysis is a key part of the requirements in the specifications and regulations. As an example, it establishes the critical configurations and areas of the envelope where flutter may be a concern. Testing is used to validate the analytical work. However, some specifications and regulations also directly require testing as a separate action, beyond the purpose of validating the analytical work. For instance, ground vibration testing to determine the natural frequencies and mode shapes, while a vital part of analytical validation, is also an independent requirement in some regulations. Similarly, flight flutter testing is a specific required demonstration that the aircraft is free from aeroelastic instabilities and meets the damping requirements. After the initial design is completed, there will be on-going modifications and configurations to address such as stores arrangements on military aircraft. Combinations of analysis and testing will, therefore, be a continual, integrated process to verify the design; see Fig. 5.18.

5.8.1 Ground-Vibration Tests

The purpose of structural dynamics experiments on the ground is to validate the frequencies and mode shapes of a clean airplane or important airplane configurations. To accomplish this, the airplane is equipped with strain gages and accelerometers at the roots and tips of the wings, of the horizontal and vertical tails, and of the airplane nose. The airplane is placed on soft supports to mimic the airplanes free-free structural dynamics. Vertical actuators (i.e., shakers) are used at the tips of the wing and horizontal tail; both vertical and side shakers may be used at the tips of the nose

---

[3] Rusak (2011), private comm.

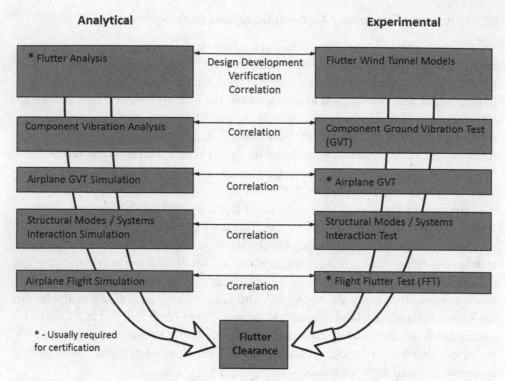

Figure 5.18. Schematic of interaction between analysis and test in certification; from Nieder-meyer (2014), Federal Aviation Administration, used with permission

and vertical tail. There is a variety of signal analysis methods to identify natural frequencies, mode shapes, and structural damping from the measurements. Generally, the actuators have a bandwidth up to 30 Hz, and a sweep of actuation frequencies is first conducted from 0.1 to 30 Hz to identify the symmetric and antisymmetric modes in this range. Classical techniques, such as Fast Fourier Transformation (FFT) and Power Spectral Density (PSD), are used for spectral analysis of unsteady elastic deformation signals to identify the natural frequencies. At this point, we must continue to study details of the dynamic response at the natural bending and torsional frequencies of interest for flutter or other aeroelastic phenomena. For each mode, this entails the following:

1. Induce oscillatory motion of the mode at a certain natural frequency, measure the response, and perform an FFT analysis to identify the resonance frequency and structural damping of that mode.
2. Induce a step-function command from oscillatory motion to zero, measure the decay rate, and infer the structural damping of that mode.
3. Induce an impulsive function, measure the decay rate, and infer structural damping of the mode.

Now we are ready to compare experimentally measured frequencies and mode shapes with detailed finite-element predictions. Using well-established techniques, we tune our finite element model to yield frequencies and mode shapes that fit the ground-vibration test data.

5.8.2 Wind Tunnel Flutter Experiments

The design of wind tunnel models that accurately represent the flutter situations of a real airplane is a complex task—quite possibly an art! The main challenge is to compose a small-scale version of the aircraft with tuned structural dynamics and a sufficiently detailed geometry that correctly reflects the aircraft's static and unsteady aerodynamic behavior. For these tests, the model is supported by soft cables and equipped with strain gages and accelerometers at the roots and tips of wings, at the tips of horizontal and vertical tails, and on the nose. Flow turbulence is used to excite the model aeroelastic modes. FFT and PSD analysis of the various measured model deformations are used to estimate the flutter speed, frequency and severity.

The wind tunnel tests provide essential insight into the possible modes of flutter of an airplane. The tests help verify computed results as well as identify unknown aeroelastic phenomena related to the airplane configuration. However, it should be recognized that unsteady flow phenomena are strongly governed by scale, so that reduced frequency, flow Reynolds number, separation between vortices, and interactions between shockwaves and boundary layers may not be correctly represented by small-scale models. In such experiments, results may lead to an inaccurate prediction of flutter occurrence in the full-scale airplane. In addition, testing of small-scale models in the wind tunnel provides benchmark cases for improving the computational models and tuning the unsteady aerodynamics analysis codes.

5.8.3 Ground Roll (Taxi) and Flight Tests

A special experimental aircraft is equipped with the capability of making realtime measurements of the amount of fuel, airspeed, Mach number, altitude, load factors, and control surface deflections. The aircraft is also instrumented as in the ground-vibration tests with strain gages and accelerometers. Special actuators are included to operate the ailerons and elevators over a range of frequencies. A sweep of actuation frequencies is first conducted to identify important modes. In addition, at certain frequencies, responses to step and impulse commands are measured.

Ground Roll (Taxi) Measurements. Aircraft ground roll (taxi) provides the first insight into an airplane's aeroelastic response. The relatively rough runway excites the airplane's structural modes. In addition, here we conduct a sweep of frequencies and FFT-PSD analyses of measurements and determine whether results match with analysis predictions of aeroelastic behavior at near zero speeds. If they do not, we must stop the test to correct and/or adjust the computational model until agreement is found, and then flutter predictions are reevaluated. Only when results do agree, may we then proceed to take off. These tests are not required directly by regulations but are an important part of technical development.

Flight Tests. In this step, we take off and fly at the lowest speed at low altitude and in level flight. We measure the airplane's response to air turbulence and conduct FFT and PSD analyses. We determine whether these results match our analytical predictions for the tested speed. If they do not, we must again stop the test to correct and/or adjust the computational model until we find agreement. When they do agree, we then may proceed to activate the actuators for an impulse command. Next, we determine whether there is sufficient damping. If there is, we conduct a sweep of frequencies and conduct FFT and PSD analyses of frequencies and damping. If these results match the analysis, then we may (cautiously!) activate actuators at the calculated flutter frequency and conduct FFT and PSD analyses of the response. When the damping measurements match theoretical predictions, then we activate actuators for a step command. Next, we determine whether the response matches the analysis. When it does, we then collect and store data in the form of U-g-g' diagrams. Only if and when there is reasonable agreement with analyses we proceed cautiously to perform maneuvers at various load factors at the same speed and altitude. During each maneuver, we activate the actuators for an impulse command to see whether there is sufficient damping. If there is, we move on to increase the load factor until the complete set of specified load factors within the flight envelope is tested.

If our computed predictions are in agreement with the results obtained at any stage, only then is it safe to go to a higher speed (e.g., 25 knots faster) at the same altitude. At this point we repeat all of the steps, collecting and storing data in the form of U-g-g' diagrams. We systematically and cautiously increase the speed up to its maximum, checking at every increment to ensure that our analysis is valid. Similarly, we systematically increase altitude to its maximum and repeat the regimen. The testing pattern may follow many other routes based on critical flutter modes. We stop (i.e., reduce speed to the previous safe speed) immediately whenever any one of the following happens:

1. A modal damping coefficient g decreases below the level of damping required by regulations (3% for both civil and military aircraft worldwide).
2. Oscillations in at least one measurement diverge and grow beyond preapproved limits.
3. The dominant frequency deviates from its predicted value.

Thus, it is observed that the analysis of airplane flutter is strongly based on theoretical studies. The theory is the work tool for analysis and decisions about critical configurations and flight conditions. Ground vibration experiments are used to tune the structural dynamics analysis to yield accurate structural modes, and wind tunnel experiments to tune the unsteady aerodynamics code. Flutter flight tests are extremely dangerous. Real-time measurements and various actuation techniques are used to estimate the damping of the airplane at various flight altitudes, speeds, and load factors and move from one point to another with much caution. Analysis, ground experiments and flight tests always go together to provide full clearance for flight without flutter problems.

5.9 Epilogue

In this chapter, we consider the general problem of lifting-surface flutter. Several types of flutter analysis were presented, including the p method, classical flutter analysis, k method, and p-k method. The application of classical flutter analysis to discrete one- and two-degree-of-freedom wind-tunnel models was presented. Students were exposed to Theodorsen's unsteady thin-airfoil theory along with the more modern finite-state thin-airfoil theory of Peters et al. Application of the assumed-modes method to construct a flutter analysis of a flexible wing was demonstrated as well. The important parameters of the flutter problem were discussed, along with current design practice, flight testing, and certification. With a good understanding of the material presented herein, students should be sufficiently equipped to apply these fundamentals to the design of flight vehicles.

Moreover, with appropriate graduate-level studies well beyond the scope of material presented herein, students will be able to conduct research in the exciting field of aeroelasticity. Current research topics are quite diverse. With the increased sophistication of controls technology, it has become more common to attack flutter problems by active control of flaps or other flight-control surfaces. These so-called flutter-suppression systems provide alternatives to costly design changes. One type of system for which flutter-suppression systems are an excellent choice is a military aircraft that must carry weapons as stores. These aircraft must be free of flutter within their flight envelope for different configurations, sometimes many different configurations. At times, avoidance of flutter by design changes is simply beyond the capability of designers for such complex systems. There is also a body of research to determine in flight when a flutter boundary is being approached. This could be of great value for situations in which damage had altered the properties of the aircraft structure—perhaps unknown to the pilot—thus shifting the flutter (or divergence) boundary and making the aircraft unsafe to operate within its original flight envelope. Other current problems of interest to aeroelasticians include improved analysis methodology for prediction of flutter, gust response, and limit-cycle oscillations; design of control systems to improve gust response and limit-cycle oscillations; and incorporation of aeroelastic analyses at an earlier stage of aircraft design.

Problems

1. Compute the flutter speed for the incompressible, one-degree-of-freedom flutter problem with

$$m_\theta = \frac{i-2}{k} - 10i$$

$$I_P = 50\pi\rho_\infty b^4 \qquad \omega_\theta = 10 \text{ Hz} \qquad b = 0.5 \text{ ft}$$

Answer: $U_F = 405.6$ ft/sec

2. According to Theodorsen's theory, the circulatory lift is proportional to a quantity that for simple harmonic motion can be shown to be equal to the effective

angle of attack given by

$$\alpha = C(k) \left[\theta + \frac{h}{U} + \frac{b}{U} \left(\frac{1}{2} - a \right) \dot{\theta} \right]$$

For $a = -1/2$ and simple harmonic motion such that $\bar{\theta} = 1$ and $\bar{h} = bz[\cos(\phi) + i \sin(\phi)]$, find the amplitude and phase of α relative to θ and plot α as a function of time for five periods for the following four cases:
(a) $z = 0.1; \phi = 0°; k = 0.01$
(b) $z = 0.1; \phi = 0°; k = 1.0$
(c) $z = 0.1; \phi = 90°; k = 1.0$
(d) $z = 0.5; \phi = 90°; k = 1.0$
Comment on the behavior of α for increasing k, changing the phase angle from 0 to 90 degrees, and increasing the plunge magnitude. Approximate Theodorsen's function as

$$C(k) = \frac{0.01365 + 0.2808ik - \frac{k^2}{2}}{0.01365 + 0.3455ik - k^2}$$

Answer: (a) amplitude: 0.9931; phase lag: 2.01°
Answer: (b) amplitude: 0.7988; phase lead: 37.0°
Answer: (c) amplitude: 0.7229; phase lead: 37.3°
Answer: (d) amplitude: 0.6008; phase lead: 52.7°

3. Show that the coefficients used in a classical flutter analysis, if based on Theodorsen's theory, are

$$\ell_h = 1 - \frac{2i\,C(k)}{k}$$

$$\ell_\theta = -a - \frac{i}{k} - \frac{2C(k)}{k^2} - \frac{2i\left(\frac{1}{2} - a\right)C(k)}{k}$$

$$m_h = -a + \frac{2i\left(\frac{1}{2} + a\right)C(k)}{k}$$

$$m_\theta = \frac{1}{8} + a^2 - \frac{i\left(\frac{1}{2} - a\right)}{k} + \frac{2\left(\frac{1}{2} + a\right)C(k)}{k^2} + \frac{2i\left(\frac{1}{4} - a^2\right)C(k)}{k}$$

4. Consider a two-dimensional rigid wing in incompressible flow with freestream speed U and pivoted about the leading edge. The pitch motion is spring-restrained with spring constant $k_\theta = I_P \omega_\theta^2$. Use the exact $C(k)$ and
(a) determine the flutter speed and flutter frequency for $I_P = 2{,}500\pi \rho_\infty b^4$
(b) determine the minimum possible flutter speed and flutter frequency
Answer: (a) $U_F = 28.2279\, b\omega_\theta$ and $\omega_F = 1.13879\,\omega_\theta$; (b) $U_F = 24.7877\, b\omega_\theta$ and $\omega_F = \omega_\theta$

5. Consider an incompressible, two-degree-of-freedom flutter problem in which $a = -1/5, e = -1/10, \mu = 20, r^2 = 6/25,$ and $\sigma = 2/5$. Compute the flutter speed

and the flutter frequency using the classical flutter approach. For the aerodynamic coefficients, use those of Theodorsen's theory with $C(k)$ approximated as in Problem 2.

Answer: $U_F = 2.170\, b\omega_\theta$ and $\omega_F = 0.6443\, \omega_\theta$

6. Consider an incompressible, two-degree-of-freedom flutter problem in which $a = -1/5$, $\mu = 3$, and $r = 1/2$. Compute the flutter speed and flutter frequency for two cases $x_\theta = e - a = 1/5$ and $x_\theta = e - a = 1/10$, and let $\sigma = 0.2, 0.4, 0.6, 0.8,$ and 1.0. Use the classical flutter approach and, for the aerodynamic coefficients, use those of Theodorsen's theory with $C(k)$ approximated as in Problem 2. Compare with the results in Fig. 5.15.

7. Set up the complete set of equations for flutter analysis by the p method using the unsteady-aerodynamic theory of Peters et al. (1995), nondimensionalizing Eqs. (5.106), and redefining λ_i as $b\omega_\theta\lambda_i$.

8. Write a computer program using MATLAB™ or Mathematica™ to set up the solution of the equations derived in Problem 7.

9. Using the computer program written in Problem 8, solve for the dimensionless flutter speed and flutter frequency for an incompressible, two-degree-of-freedom flutter problem in which $a = -1/3$, $e = -1/10$, $\mu = 50$, $r = 2/5$, and $\sigma = 2/5$. Answer: $U_F = 2.807\, b\omega_\theta$ and $\omega_F = 0.5952\, \omega_\theta$

10. Write a computer program using MATLAB™ or Mathematica™ to set up the solution of a two-degree-of-freedom flutter problem using the k method.

11. Use the computer program written in Problem 10 to solve a flutter problem in which $a = -1/5$, $e = -1/10$, $\mu = 20$, $r^2 = 6/25$, and $\sigma = 2/5$. Plot the values of $\omega_{1,2}/\omega_\theta$ and g versus $U/(b\omega_\theta)$ and compare your results with the quantities plotted in Figs. 5.12 and 5.13. Noting how the quantities plotted in these two sets of figures are different, comment on the similarities and differences observed in these plots and why those differences are there. Finally, explain why your predicted flutter speed is the same as that determined by the classical method.

Answer: See Figs. 5.19 and 5.20.

12. Show that the flutter determinant for the p-k method applied to the typical section using Theodorsen aerodynamics can be expressed as

$$\begin{vmatrix} p^2 + \frac{\sigma^2}{V^2} - \frac{k^2}{\mu} + \frac{2ikC(k)}{\mu} & \frac{p^2\mu x_\theta + k(i+ak) + [2+ik(1-2a)]C(k)}{\mu} \\ \frac{p^2\mu x_\theta + ak^2 - ik(1+2a)C(k)}{\mu} & \frac{8\mu r^2\left(p^2 + \frac{1}{V^2}\right) + 4i(1+2a)[2i - k(1-2a)]C(k) - k[k - 4i + 8a(i+ak)]}{8\mu} \end{vmatrix}$$

13. Write a computer program using MATLAB™ or Mathematica™ to set up the solution of a two-degree-of-freedom flutter problem using the p-k method and Theodorsen aerodynamics.

14. Use the computer program written in Problem 13 to solve a flutter problem in which $a = -1/5$, $e = -1/10$, $\mu = 20$, $r^2 = 6/25$, and $\sigma = 2/5$. Plot the values of

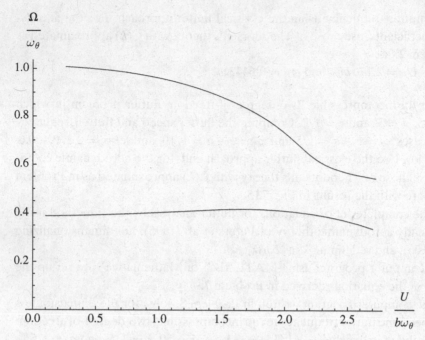

Figure 5.19. Plot of $\omega_{1,2}/\omega_\theta$ versus $U/(b\omega_\theta)$ using the k method and Theodorsen aerodynamics with $a = -1/5$, $e = -1/10$, $\mu = 20$, $r^2 = 6/25$, and $\sigma = 2/5$

the estimates of $\Omega_{1,2}/\omega_\theta$ and $\Gamma_{1,2}/\omega_\theta$ versus $U/(b\omega_\theta)$ and compare your results with the quantities plotted in Figs. 5.12 and 5.13. Explain why the estimated damping from the p-k method sometimes differs from that of the p method. Answer: See Figs. 5.21 and 5.22.

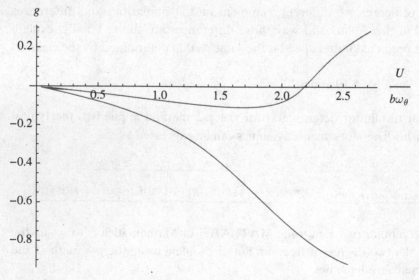

Figure 5.20. Plot of g versus $U/(b\omega_\theta)$ using the k method and Theodorsen aerodynamics with $a = -1/5$, $e = -1/10$, $\mu = 20$, $r^2 = 6/25$, and $\sigma = 2/5$

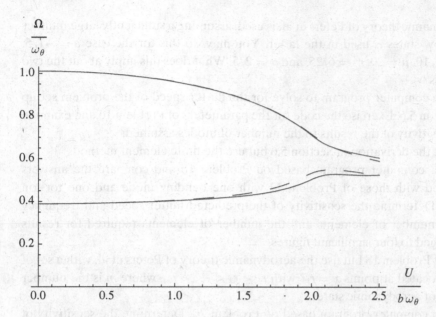

Figure 5.21. Plot of estimated value of $\Omega_{1,2}/\omega_\theta$ versus $U/(b\omega_\theta)$ using the p-k method with Theodorsen aerodynamics (dashed lines) and the p method with the aerodynamics of Peters et al. (solid lines) for $a = -1/5$, $e = -1/10$, $\mu = 20$, $r^2 = 6/25$, and $\sigma = 2/5$

15. Write a computer program using MATLAB™ or Mathematica™ to set up the solution of a two-degree-of-freedom flutter problem using the p-k method and the aerodynamics of Peters et al.

16. Using the computer programs of Problems 13 and 15, show that the p-k method yields the same results regardless of whether Theodorsen's theory or the

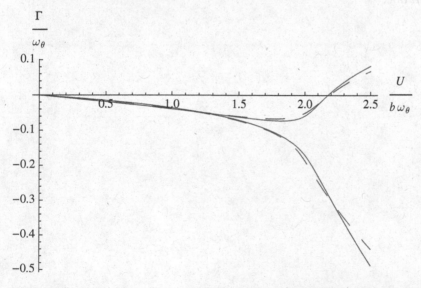

Figure 5.22. Plot of estimated value of $\Gamma_{1,2}/\omega_\theta$ versus $U/(b\omega_\theta)$ using the p-k method with Theodorsen aerodynamics (dashed lines) and the p method with the aerodynamics of Peters et al. (solid lines) for $a = -1/5$, $e = -1/10$, $\mu = 20$, $r^2 = 6/25$, and $\sigma = 2/5$

aerodynamic theory of Peters et al. is used, assuming a sufficiently large number of inflow states is used in the latter. You may do this for the case $a = -1/5$, $e = -1/10$, $\mu = 20$, $r^2 = 6/25$, and $\sigma = 2/5$. What does this imply about the two theories?

17. Write a computer program to solve for the flutter speed of the problem set up in Section 5.6. Exercise the code for the parameters of Problem 16 and examine the sensitivity of the results to the number of modes assumed.

18. Repeat the derivation in Section 5.6 but use the finite element method.

19. Write a computer program based on Problem 18 and compare the answers obtained with those of Problem 17 with one bending mode and one torsion mode. Determine the sensitivity of the predicted flutter speed and frequency to the number of elements and the number of elements required for results converged to four significant figures.

20. Repeat Problem 18 but use the aerodynamic theory of Peters et al. with a set of states located at points $x = r_i \ell$ with $r_1 < r_2 < \ldots < r_m$, where m is the number of sets of aerodynamic states.

21. Write a computer program based on Problem 20. Determine the sensitivity of the results to the number and values of parameters r_i, in which the latter are equally spaced along the span.

22. Comparing Eqs. (5.11) and (5.71), find an expression for $[\mathcal{A}(p)]$ in terms of matrices $[a]$, $[b]$, $[c]$, $[d]$, $[A]$, and $[E]$ for the special case when the only degrees of freedom in the column matrix $\bar{\xi}$ are $\bar{h}/b$ and $\bar{\theta}$, and the unsteady-aerodynamic theory is based on the theory of Peters et al. with six states. Assuming simple harmonic motion, extract an approximation for $C(k)$ from these equations. Compare the real and imaginary parts of $C(k)$ with those from the approximation.

APPENDIX A

Lagrange's Equations

A.1 Introduction

When we wish to use Newton's laws to write the equations of motion of a particle or a system of particles, we must be careful to include all the forces of the system. The Lagrangean form of the equations of motion that we derive herein has the advantage that we can ignore all forces that do no work (e.g., forces at frictionless pins, forces at a point of rolling contact, forces at frictionless guides, and forces in inextensible connections). In the case of conservative systems (i.e., systems for which the total energy remains constant), the Lagrangean method gives us an automatic procedure for obtaining the equations of motion provided only that we can write the kinetic and potential energies of the system.

A.2 Degrees of Freedom

Before proceeding to develop the Lagrange equations, we must characterize our dynamical systems in a systematic way. The most important property of this sort for our present purpose is the number of independent coordinates that we must know to completely specify the position or configuration of our system. We say that a system has n degrees of freedom if exactly n coordinates serve to completely define its configuration.

EXAMPLE 1 A free particle in space has three degrees of freedom because we must know three coordinates—x, y, z, for example – to locate it.

EXAMPLE 2 A wheel that rolls without slipping on a straight track has one degree of freedom because either the distance from some base point or the total angle of rotation will enable us to locate it completely.

A.3 Generalized Coordinates

We usually think of coordinates as lengths or angles. However, any set of parameters that enables us to uniquely specify the configuration of the system can serve as

231

coordinates. When we generalize the meaning of the term in this manner, we call these new quantities "generalized coordinates."

EXAMPLE 3 Consider a bar rotating in a plane about a point O. The angle of rotation with respect to some base line is suggested as an obvious coordinate for specifying the position of the bar. However, the area swept over by the bar would do equally well and therefore could be used as a generalized coordinate.

If a system has n degrees of freedom, then n generalized coordinates are necessary and sufficient to determine the configuration.

A.4 Lagrange's Equations

In deriving these equations, we consider systems having two degrees of freedom and hence are completely defined by two generalized coordinates q_1 and q_2. However, the results are easily extended to systems with any number of degrees of freedom.

Suppose our system consists of n particles. For each particle, we can write by Newton's second law

$$M_i \ddot{x}_i = X_i$$
$$M_i \ddot{y}_i = Y_i \tag{A.1}$$
$$M_i \ddot{z}_i = Z_i$$

where x_i, y_i, and z_i are the rectangular Cartesian coordinates of the ith particle; M_i is the mass; and X_i, Y_i, and Z_i are the resultants of all forces acting on it in the x, y, and z directions, respectively.

If we multiply both sides of Eqs. (A.1) by δx_i, δy_i, and δz_i, respectively, and add the equations, we have

$$M_i \left(\ddot{x}_i \delta x_i + \ddot{y}_i \delta y_i + \ddot{z}_i \delta z_i \right) = X_i \delta x_i + Y_i \delta y_i + Z_i \delta z_i \tag{A.2}$$

The right-hand side of this equation represents the work done by all of the forces acting on the ith particle during the virtual displacements δx_i, δy_i, and δz_i. Hence, forces that do no work do not contribute to the right-hand side of Eq. (A.2) and may be omitted from the equation. To obtain the corresponding equation for the entire system, we sum both sides of Eq. (A.2) for all particles. Thus

$$\sum_{i=1}^{n} M_i \left(\ddot{x}_i \delta x_i + \ddot{y}_i \delta y_i + \ddot{z}_i \delta z_i \right) = \sum_{i=1}^{n} \left(X_i \delta x_i + Y_i \delta y_i + Z_i \delta z_i \right) \tag{A.3}$$

Now, because our system is completely located in space if we know the two generalized coordinates q_1 and q_2, we must be able to write x_i, y_i, and z_i as well as their increments δx_i, δy_i, and δz_i as functions of q_1 and q_2. Hence

$$x_i = x_i(q_1, q_2)$$
$$y_i = y_i(q_1, q_2) \qquad \text{(A.4)}$$
$$z_i = z_i(q_1, q_2)$$

Differentiating Eq. (A.4) with respect to time gives

$$\dot{x}_i = \frac{\partial x_i}{\partial q_1}\dot{q}_1 + \frac{\partial x_i}{\partial q_2}\dot{q}_2$$
$$\dot{y}_i = \frac{\partial y_i}{\partial q_1}\dot{q}_1 + \frac{\partial y_i}{\partial q_2}\dot{q}_2 \qquad \text{(A.5)}$$
$$\dot{z}_i = \frac{\partial z_i}{\partial q_1}\dot{q}_1 + \frac{\partial z_i}{\partial q_2}\dot{q}_2$$

Similarly

$$\delta x_i = \frac{\partial x_i}{\partial q_1}\delta q_1 + \frac{\partial x_i}{\partial q_2}\delta q_2$$
$$\delta y_i = \frac{\partial y_i}{\partial q_1}\delta q_1 + \frac{\partial y_i}{\partial q_2}\delta q_2 \qquad \text{(A.6)}$$
$$\delta z_i = \frac{\partial z_i}{\partial q_1}\delta q_1 + \frac{\partial z_i}{\partial q_2}\delta q_2$$

If we substitute these into Eq. (A.3) and rearrange the terms, we obtain

$$
\begin{aligned}
\sum_{i=1}^{n} &\left[M_i \left(\ddot{x}_i \frac{\partial x_i}{\partial q_1} + \ddot{y}_i \frac{\partial y_i}{\partial q_1} + \ddot{z}_i \frac{\partial z_i}{\partial q_1} \right) \delta q_1 \right. \\
&\left. + M_i \left(\ddot{x}_i \frac{\partial x_i}{\partial q_2} + \ddot{y}_i \frac{\partial y_i}{\partial q_2} + \ddot{z}_i \frac{\partial z_i}{\partial q_2} \right) \delta q_2 \right] \\
&= \sum_{i=1}^{n} \left[\left(X_i \frac{\partial x_i}{\partial q_1} + Y_i \frac{\partial y_i}{\partial q_1} + Z_i \frac{\partial z_i}{\partial q_1} \right) \delta q_1 \right. \\
&\left. + \left(X_i \frac{\partial x_i}{\partial q_2} + Y_i \frac{\partial y_i}{\partial q_2} + Z_i \frac{\partial z_i}{\partial q_2} \right) \delta q_2 \right]
\end{aligned}
\qquad \text{(A.7)}
$$

From Eq. (A.5), we conclude that because x_i, y_i, and z_i are functions of q_1 and q_2 but not of $\dot{q}_1$ and $\dot{q}_2$, then

$$
\frac{\partial \dot{x}_i}{\partial \dot{q}_1} = \frac{\partial x_i}{\partial q_1} \qquad \frac{\partial \dot{x}_i}{\partial \dot{q}_2} = \frac{\partial x_i}{\partial q_2}
$$
$$
\frac{\partial \dot{y}_i}{\partial \dot{q}_1} = \frac{\partial y_i}{\partial q_1} \qquad \frac{\partial \dot{y}_i}{\partial \dot{q}_2} = \frac{\partial y_i}{\partial q_2} \qquad \text{(A.8)}
$$
$$
\frac{\partial \dot{z}_i}{\partial \dot{q}_1} = \frac{\partial z_i}{\partial q_1} \qquad \frac{\partial \dot{z}_i}{\partial \dot{q}_2} = \frac{\partial z_i}{\partial q_2}
$$

We substitute these relationships into the left-hand side of Eq. (A.7) to obtain

$$
\sum_{i=1}^{n} \left[M_i \left(\ddot{x}_i \frac{\partial \dot{x}_i}{\partial \dot{q}_1} + \ddot{y}_i \frac{\partial \dot{y}_i}{\partial \dot{q}_1} + \ddot{z}_i \frac{\partial \dot{z}_i}{\partial \dot{q}_1} \right) \delta q_1 \right.
$$
$$
\left. + M_i \left(\ddot{x}_i \frac{\partial \dot{x}_i}{\partial \dot{q}_2} + \ddot{y}_i \frac{\partial \dot{y}_i}{\partial \dot{q}_2} + \ddot{z}_i \frac{\partial \dot{z}_i}{\partial \dot{q}_2} \right) \delta q_2 \right]
$$
$$
= \sum_{i=1}^{n} \left[\left(X_i \frac{\partial x_i}{\partial q_1} + Y_i \frac{\partial y_i}{\partial q_1} + Z_i \frac{\partial z_i}{\partial q_1} \right) \delta q_1 \right.
$$
$$
\left. + \left(X_i \frac{\partial x_i}{\partial q_2} + Y_i \frac{\partial y_i}{\partial q_2} + Z_i \frac{\partial z_i}{\partial q_2} \right) \delta q_2 \right]
$$

(A.9)

Now, let us shift our attack on the problem and consider the kinetic energy of the system. This is

$$
K = \frac{1}{2} \sum_{i=1}^{n} M_i \left(\dot{x}_i^2 + \dot{y}_i^2 + \dot{z}_i^2 \right)
$$

(A.10)

Now calculate $\frac{\partial K}{\partial \dot{q}_1}$ and $\frac{\partial K}{\partial \dot{q}_2}$ to obtain

$$
\frac{\partial K}{\partial \dot{q}_1} = \sum_{i=1}^{n} M_i \left(\dot{x}_i \frac{\partial \dot{x}_i}{\partial \dot{q}_1} + \dot{y}_i \frac{\partial \dot{y}_i}{\partial \dot{q}_1} + \dot{z}_i \frac{\partial \dot{z}_i}{\partial \dot{q}_1} \right)
$$

(A.11)

$$
\frac{\partial K}{\partial q_1} = \sum_{i=1}^{n} M_i \left(\dot{x}_i \frac{\partial \dot{x}_i}{\partial q_1} + \dot{y}_i \frac{\partial \dot{y}_i}{\partial q_1} + \dot{z}_i \frac{\partial \dot{z}_i}{\partial q_1} \right)
$$

(A.12)

We next calculate the time derivative of $\frac{\partial x_i}{\partial q_1}$, for which the chain rule gives

$$
\frac{d}{dt} \left(\frac{\partial x_i}{\partial q_1} \right) = \frac{\partial^2 x_i}{\partial q_1^2} \dot{q}_1 + \frac{\partial^2 x_i}{\partial q_1 \partial q_2} \dot{q}_2
$$
$$
= \frac{\partial}{\partial q_1} \left(\frac{\partial x_i}{\partial q_1} \dot{q}_1 + \frac{\partial x_i}{\partial q_2} \dot{q}_2 \right)
$$
$$
= \frac{\partial}{\partial q_1} \left(\dot{x}_i \right) = \frac{\partial \dot{x}_i}{\partial q_1}
$$

(A.13)

Because from Eq. (A.8) we have

$$
\frac{\partial \dot{x}_i}{\partial \dot{q}_1} = \frac{\partial x_i}{\partial q_1}
$$

(A.14)

we conclude from Eq. (A.13) that

$$
\frac{d}{dt} \left(\frac{\partial x_i}{\partial q_1} \right) = \frac{\partial \dot{x}_i}{\partial q_1}
$$

(A.15)

The following relationships can be proven in a similar manner:

$$\frac{d}{dt}\left(\frac{\partial y_i}{\partial q_1}\right) = \frac{\partial \dot{y}_i}{\partial q_1}$$

$$\frac{d}{dt}\left(\frac{\partial z_i}{\partial q_1}\right) = \frac{\partial \dot{z}_i}{\partial q_1}$$

(A.16)

Now let us use Eqs. (A.11), (A.12), (A.15), and (A.16) to calculate the function

$$\frac{d}{dt}\left(\frac{\partial K}{\partial \dot{q}_1}\right) - \frac{\partial K}{\partial q_1}$$

(A.17)

for which the result is

$$\frac{d}{dt}\left(\frac{\partial K}{\partial \dot{q}_1}\right) - \frac{\partial K}{\partial q_1} = \sum_{i=1}^{n} M_i \left(\ddot{x}_i \frac{\partial \dot{x}_i}{\partial \dot{q}_1} + \ddot{y}_i \frac{\partial \dot{y}_i}{\partial \dot{q}_1} + \ddot{z}_i \frac{\partial \dot{z}_i}{\partial \dot{q}_1}\right)$$

$$+ \sum_{i=1}^{n} M_i \left[\dot{x}_i \frac{d}{dt}\left(\frac{\partial \dot{x}_i}{\partial \dot{q}_1}\right) + \dot{y}_i \frac{d}{dt}\left(\frac{\partial \dot{y}_i}{\partial \dot{q}_1}\right) + \dot{z}_i \frac{d}{dt}\left(\frac{\partial \dot{z}_i}{\partial \dot{q}_1}\right)\right]$$

$$- \sum_{i=1}^{n} M_i \left(\dot{x}_i \frac{\partial \dot{x}_i}{\partial q_1} + \dot{y}_i \frac{\partial \dot{y}_i}{\partial q_1} + \dot{z}_i \frac{\partial \dot{z}_i}{\partial q_1}\right)$$

(A.18)

From Eqs. (A.15) and (A.16), the second and third terms on the right-hand side of Eq. (A.18) are equal and thus cancel, leaving

$$\frac{d}{dt}\left(\frac{\partial K}{\partial \dot{q}_1}\right) - \frac{\partial K}{\partial q_1} = \sum_{i=1}^{n} M_i \left(\ddot{x}_i \frac{\partial \dot{x}_i}{\partial \dot{q}_1} + \ddot{y}_i \frac{\partial \dot{y}_i}{\partial \dot{q}_1} + \ddot{z}_i \frac{\partial \dot{z}_i}{\partial \dot{q}_1}\right)$$

(A.19)

A similar relationship holds for partial derivatives of K with respect to q_2 and $\dot{q}_2$. Hence, Eq. (A.9) can be written

$$\left[\frac{d}{dt}\left(\frac{\partial K}{\partial \dot{q}_1}\right) - \frac{\partial K}{\partial q_1}\right]\delta q_1 + \left[\frac{d}{dt}\left(\frac{\partial K}{\partial \dot{q}_2}\right) - \frac{\partial K}{\partial q_2}\right]\delta q_2$$

$$= \sum_{i=1}^{n}\left(X_i \frac{\partial x_i}{\partial q_1} + Y_i \frac{\partial y_i}{\partial q_1} + Z_i \frac{\partial z_i}{\partial q_1}\right)\delta q_1$$

(A.20)

$$+ \sum_{i=1}^{n}\left(X_i \frac{\partial x_i}{\partial q_2} + Y_i \frac{\partial y_i}{\partial q_2} + Z_i \frac{\partial z_i}{\partial q_2}\right)\delta q_2$$

Because q_1 and q_2 are independent coordinates, they can be varied arbitrarily. Hence, we can conclude that

$$\frac{d}{dt}\left(\frac{\partial K}{\partial \dot{q}_1}\right) - \frac{\partial K}{\partial q_1} = \sum_{i=1}^{n}\left(X_i \frac{\partial x_i}{\partial q_1} + Y_i \frac{\partial y_i}{\partial q_1} + Z_i \frac{\partial z_i}{\partial q_1}\right)$$

$$\frac{d}{dt}\left(\frac{\partial K}{\partial \dot{q}_2}\right) - \frac{\partial K}{\partial q_2} = \sum_{i=1}^{n}\left(X_i \frac{\partial x_i}{\partial q_2} + Y_i \frac{\partial y_i}{\partial q_2} + Z_i \frac{\partial z_i}{\partial q_2}\right)$$

(A.21)

The right-hand side of Eq. (A.20) is the work done by all of the forces on the system when the coordinates of the ith particle undergo the small displacement δx_i, δy_i, and δz_i due to changes δq_1 and δq_2 in the generalized coordinates q_1 and q_2. The coefficients of δq_1 and δq_2 are known as the generalized forces Q_1 and Q_2 because they are the quantities by which the variations of the generalized coordinates must be multiplied to calculate the virtual work done by all the forces acting on the system. Hence

$$Q_1 = \sum_{i=1}^{n} \left(X_i \frac{\partial x_i}{\partial q_1} + Y_i \frac{\partial y_i}{\partial q_1} + Z_i \frac{\partial z_i}{\partial q_1} \right)$$

$$Q_2 = \sum_{i=1}^{n} \left(X_i \frac{\partial x_i}{\partial q_2} + Y_i \frac{\partial y_i}{\partial q_2} + Z_i \frac{\partial z_i}{\partial q_2} \right)$$

(A.22)

and Eqs. (A.21) can be written

$$\frac{d}{dt} \left(\frac{\partial K}{\partial \dot{q}_1} \right) - \frac{\partial K}{\partial q_1} = Q_1$$

$$\frac{d}{dt} \left(\frac{\partial K}{\partial \dot{q}_2} \right) - \frac{\partial K}{\partial q_2} = Q_2$$

(A.23)

This is one form of Lagrange's equations of motion. They apply to any system that is completely described by two and only two generalized coordinates, whether or not the system is conservative. It can be shown by slightly more extended calculation that they apply to systems of any finite number of degrees of freedom.

A.5 Lagrange's Equations for Conservative Systems

If a system is conservative, the work done by the forces can be calculated from the potential energy P. We define the change in potential energy during a small displacement as the negative of the work done by the forces of the system during the displacement. Because $Q_1 \delta q_1 + Q_2 \delta q_2$ is the work done by the forces, we have

$$\delta P = -Q_1 \delta q_1 - Q_2 \delta q_2$$

(A.24)

We have emphasized that q_1 and q_2 are independent and, hence, can be varied arbitrarily. If $\delta q_2 = 0$, we have $\delta P = -Q_1 \delta q_1$ so that

$$Q_1 = -\frac{\partial P}{\partial q_1}$$

(A.25)

Similarly, it can be seen that

$$Q_2 = -\frac{\partial P}{\partial q_2}$$

(A.26)

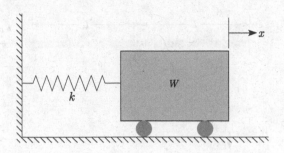

Figure A.1. Schematic for the mechanical system
of Example 5

Replacing Q_1 and Q_2 in Eqs. (A.23) by these expressions, we have

$$\frac{d}{dt}\left(\frac{\partial K}{\partial \dot{q}_1}\right) - \frac{\partial K}{\partial q_1} + \frac{\partial P}{\partial q_1} = 0$$

$$\frac{d}{dt}\left(\frac{\partial K}{\partial \dot{q}_2}\right) - \frac{\partial K}{\partial q_2} + \frac{\partial P}{\partial q_2} = 0 \qquad (A.27)$$

These are Lagrange's equations of motion for a conservative system. As before, they hold for systems of any finite number of degrees of freedom.

EXAMPLE 4 Find the equations of motion of a particle of weight W moving in space under the force of gravity.

Solution: We need three coordinates to describe the position of the particle and can therefore take x, y, and z as the generalized coordinates. Taking x and y in the horizontal plane and z vertically upward with the origin at the earth's surface and taking the origin as the zero position for potential energy, we obtain

$$K = \frac{W}{2g}\left(\dot{x}^2 + \dot{y}^2 + \dot{z}^2\right) \qquad P = Wz$$

$$\frac{\partial K}{\partial \dot{x}} = \frac{W}{g}\dot{x} \qquad \frac{\partial K}{\partial \dot{y}} = \frac{W}{g}\dot{y} \qquad \frac{\partial K}{\partial \dot{z}} = \frac{W}{g}\dot{z} \qquad \frac{\partial K}{\partial x} = \frac{\partial K}{\partial y} = \frac{\partial K}{\partial z} = 0$$

$$\frac{d}{dt}\left(\frac{\partial K}{\partial \dot{x}}\right) = \frac{W}{g}\ddot{x} \qquad \frac{d}{dt}\left(\frac{\partial K}{\partial \dot{y}}\right) = \frac{W}{g}\ddot{y} \qquad \frac{d}{dt}\left(\frac{\partial K}{\partial \dot{z}}\right) = \frac{W}{g}\ddot{z}$$

$$\frac{\partial P}{\partial x} = \frac{\partial P}{\partial y} = 0 \qquad \frac{\partial P}{\partial z} = W \qquad (A.28)$$

Hence, Lagrange's equation, Eq. (A.27), gives

$$\frac{W}{g}\ddot{x} = 0 \qquad \frac{W}{g}\ddot{y} = 0 \qquad \frac{W}{g}\ddot{z} + W = 0 \qquad (A.29)$$

Of course, these equations are more easily obtainable by the direct application of Newton's second law; this example merely illustrates the application of Lagrange's equations for a familiar problem.

EXAMPLE 5 Find the equation of motion of the sprung weight W sliding on a smooth horizontal plane (Fig. A.1).

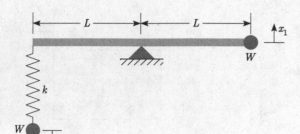

Figure A.2. Schematic for the mechanical system of Example 6

Solution: We may take x as the generalized coordinate and measure it from the equilibrium position. Then

$$K = \frac{W}{2g}\dot{x}^2 \qquad P = \frac{k}{2}x^2$$

$$\frac{\partial K}{\partial \dot{x}} = \frac{W}{g}\dot{x} \qquad \frac{d}{dt}\left(\frac{\partial K}{\partial \dot{x}}\right) = \frac{W}{g}\ddot{x} \tag{A.30}$$

$$\frac{\partial K}{\partial x} = 0 \qquad \frac{\partial P}{\partial x} = kx$$

Lagrange's equation, Eq. (A.27), gives

$$\frac{W}{g}\ddot{x} + kx = 0 \tag{A.31}$$

as the equation of motion.

EXAMPLE 6 Obtain the equations of motion for the system shown in Fig. A.2. The bar is weightless.

Solution: The coordinates x_1 and x_2 can be taken as generalized coordinates. Take as the zero datum the configuration for which the bar is horizontal and the spring is unstretched. Then

$$K = \frac{W}{2g}\dot{x}_1^2 + \frac{W}{2g}\dot{x}_2^2 \qquad P = Wx_1 - Wx_2 + \frac{1}{2}k(x_2 - x_1)^2$$

$$\frac{\partial K}{\partial \dot{x}_1} = \frac{W}{g}\dot{x}_1 \qquad \frac{\partial K}{\partial \dot{x}_2} = \frac{W}{g}\dot{x}_2 \qquad \frac{\partial K}{\partial x_1} = \frac{\partial K}{\partial x_2} = 0$$

$$\frac{d}{dt}\left(\frac{\partial K}{\partial \dot{x}_1}\right) = \frac{W}{g}\ddot{x}_1 \qquad \frac{d}{dt}\left(\frac{\partial K}{\partial \dot{x}_2}\right) = \frac{W}{g}\ddot{x}_2 \tag{A.32}$$

$$\frac{\partial P}{\partial x_1} = W - k(x_2 - x_1) \qquad \frac{\partial P}{\partial x_2} = -W + k(x_2 - x_1)$$

The Lagrange equations are

$$\frac{W}{g}\ddot{x}_1 + W - k(x_2 - x_1) = 0$$

$$\frac{W}{g}\ddot{x}_2 - W + k(x_2 - x_1) = 0 \tag{A.33}$$

This is an example of a two-degree-of-freedom conservative system.

A.6 Lagrange's Equations for Nonconservative Systems

If the system is nonconservative, then, in general, there are some forces (i.e., conservative) that are derivable from a potential function, $P(q_1, q_2, \ldots)$ and some forces (i.e., nonconservative) that are not. Those forces for which a potential function does not exist must be introduced by first determining their virtual work. The coefficient of the virtual displacement δq_i in the virtual-work expression is the generalized force, here denoted by Q_i $(i = 1, 2, \ldots)$. In this instance, it is convenient to introduce what is called the Lagrangean as

$$L = K - P \tag{A.34}$$

and write the general form of Lagrange's equations as

$$\frac{d}{dt}\left(\frac{\partial L}{\partial \dot{q}_i}\right) - \frac{\partial L}{\partial q_i} = Q_i \qquad (i = 1, 2, \ldots, n) \tag{A.35}$$

EXAMPLE 7 Rework Example 5 with a dashpot of constant c connected in parallel with the spring.

Solution: The system with a dashpot is nonconservative. Hence, we use Lagrange's equations in the form of Eq. (A.35). The kinetic and potential energies are the same as in Example 5. To calculate the Q for the dashpot force, use the definition that Q is the coefficient by which the generalized coordinates must be multiplied to obtain the work done. In any small displacement δx, the work done by the dashpot force $-c\dot{x}$ is $-c\dot{x}\,\delta x$. Hence, $-c\dot{x}$ is the generalized force associated with the dashpot. The Lagrangean is

$$L = \frac{W\dot{x}^2}{2g} - \frac{kx^2}{2} \tag{A.36}$$

and

$$Q = -c\dot{x} \tag{A.37}$$

Lagrange's equation then becomes

$$\frac{W}{g}\ddot{x} + kx = -c\dot{x} \tag{A.38}$$

References

Structural Dynamics

1. Bisplinghoff, R. L., H. Ashley, and R. L. Halfman, *Aeroelasticity*, Addison-Wesley Publishing Co., Inc., 1955.
2. Chen, Y., *Vibrations: Theoretical Methods*, Addison-Wesley Publishing Co., Inc., 1966.
3. Frazer, R. A. and Duncan, W. J., "The Flutter of Aeroplane Wings," R&M 1155, Aeronautical Research Council, August 1928.
4. Fry'ba, L., *Vibration and Solids and Structures under Moving Loads*, Noordhoff International Publishing, 1972.
5. Hodges, D. H., *Nonlinear Composite Beam Theory*, AIAA, 2006.
6. Hurty, W. C., and M. F. Rubenstein, *Dynamics of Structures*, Prentice-Hall, Inc., 1964.
7. Kalnins, A., and C. L. Dym, *Vibration: Beams, Plates and Shells*, Dowden, Hutchinson and Ross, Inc., 1976.
8. Lamb, H., *The Dynamical Theory of Sound*, 2nd Edition 1925 (reprinted by Dover).
9. Leipholz, H., *Direct Variational Methods and Eigenvalue Problems in Engineering*, Noordhoff International Publishing, 1977.
10. Lin, Y. K., *Probabilistic Theory of Structural Dynamics*, McGraw-Hill, Inc., 1967, reprinted by Robert E. Krieger Publishing Co., 1976.
11. Meirovitch, L., *Computational Methods in Structural Dynamics*, Sijthoff and Noordhoff, 1980.
12. Meirovitch, L., *Principles and Techniques of Vibrations*, Prentice-Hall, Inc., 1997.
13. Pestel, E. C., and F. A. Leckie, *Matrix Methods in Elastomechanics*, McGraw-Hill Book Co., Inc., 1963.
14. Rayleigh, J. W. S., *The Theory of Sound*, Vol. I, 1877, The MacMillan Co., 1945 (reprinted by Dover).
15. Reddy, J. N., *An Introduction to the Finite Element Method*, McGraw-Hill Book Co., Inc., 1993.
16. Roxbee Cox, H., and Pugsley, A. G., "Theory of Loss of Lateral Control Due to Wing Twisting," R&M 1506, Aeronautical Research Council, October 1932.
17. Scanlan, R. H., and R. Rosenbaum, *Introduction to the Study of Aircraft Vibration and Flutter*, The MacMillan Co., 1951 (reprinted by Dover).
18. Simitses, G. J., and D. H. Hodges, Structural Stability, Vol. 3, Pt. 13. In: *Encyclopedia of Aerospace Engineering*, Eds. R. Blockley and W. Shyy, John Wiley & Sons, 2010.
19. Snowdon, J. C., *Vibration and Shock in Damped Mechanical Systems*, John Wiley and Sons, 1968.
20. Steidel, R. F., Jr., *An Introduction to Mechanical Vibrations*, 2nd ed., John Wiley and Sons, 1979.
21. Thomson, W. T., and M. D. Dahleh, *Theory of Vibration with Applications*, 5th ed., Prentice-Hall, Inc., 1998.
22. Timoshenko, S., and D. H. Young, *Vibration Problems in Engineering*, 3rd ed., Van Nostrand Reinhold Co., 1955.

23. Tse, F. S., I. E. Morse, and R. T. Hinkle, *Mechanical Vibrations: Theory and Applications*, 2nd ed., Allyn and Bacon, Inc., 1978.
24. Zienkiewicz, O. C., and Taylor, R. L., *The Finite Element Method for Solid and Structural Mechanics*, Elsevier Butterworth-Heinemann, 2005.

Aeroelasticity

1. Abramson, H. N., *An Introduction to the Dynamics of Airplanes*, Ronald Press Co., 1958 (reprinted by Dover).
2. Anon., "Manual on Aeroelasticity," six volumes by NATO Advisory Group for Aeronautical Research and Development, 1956–1970.
3. Bisplinghoff, R. L., and H. Ashley, *Principles of Aeroelasticity*, John Wiley and Sons, Inc., 1962 (reprinted by Dover).
4. Bisplinghoff, R. L., H. Ashley, and R. L. Halfman, *Aeroelasticity*, Addison-Wesley Publishing Co., Inc., 1955.
5. Collar, A. R., "The First Fifty Years of Aeroelasticity," *Aerospace*, Vol. 5 (Paper No. 545), February 1978, pp. 12–20.
6. Diederich, F. W., and B. Budiansky, "Divergence of Swept Wings," NACA TN 1680, 1948.
7. Dowell, E. H., *Aeroelasticity of Plates and Shells*, Noordhoff International Publishing, 1975.
8. Dowell, E. H., E. F. Crawley, H. C. Curtiss, Jr., D. A. Peters, R. H. Scanlan, and F. Sisto, *A Modern Course in Aeroelasticity*, 3rd ed., Kluwer Academic Publishers, 1995.
9. Drela, M., "Transonic Low-Reynolds Number Airfoils," *Journal of Aircraft*, Vol. 29, No. 6, 1992, pp. 1106–13.
10. Freberg, C. R., and E. N. Kemler, *Aircraft Vibration and Flutter*, John Wiley and Sons, Inc., 1944.
11. Fung, Y. C., *An Introduction to the Theory of Aeroelasticity*, John Wiley and Sons, Inc., 1955 (reprinted by Dover).
12. Garrick, I. E., and W. H. Reed III, "Historical Development of Aircraft Flutter," *Journal of Aircraft*, Vol. 18, No. 11, Nov. 1981, pp. 897–912.
13. Goodman, C., "Accurate Subcritical Damping Solution of Flutter Equation Using Piecewise Aerodynamic Function," *Journal of Aircraft*, Vol. 38, No. 4, July-Aug. 2001, pp. 755–63.
14. Hassig, H. J., "An Approximate True Damping Solution of the Flutter Equation by Determinant Iteration," *Journal of Aircraft*, Vol. 8, No. 11, Nov. 1971, pp. 885–9.
15. Irwin, C. A. K., and P. R. Guyett, "The Subcritical Response and Flutter of a Swept Wing Model," Tech. Rept. 65186, Aug. 1965, Royal Aircraft Establishment, Farnborough, UK.
16. Niedermeyer, Carl: Federal Aviation Administration, private communication, 2014.
17. Noles, G. Keith: Federal Aviation Administration, private communication, 2014.
18. Peters, D. A., S. Karunamoorthy, and W.-M. Cao, "Finite State Induced-Flow Models; Part I: Two-Dimensional Thin Airfoil," *Journal of Aircraft*, Vol. 32, No. 2, Mar.-Apr. 1995, pp. 313–22.
19. Rusak, Zvi: Aeroelasticity Class Notes, Rensselaer Polytechnic Institute, Troy, New York, private communication, 2011.
20. Scanlan, R. H., and R. Rosenbaum, "Outline of an Acceptable Method of Vibration and Flutter Analysis for a Conventional Airplane," *CAA Aviation Safety Release 302*, Oct. 1948.
21. Scanlan, R. H., and R. Rosenbaum, *Introduction to the Study of Aircraft Vibration and Flutter*, The MacMillan Co., 1951 (reprinted by Dover).
22. Smith, M. J., Cesnik, C. E. S., and Hodges, D. H., "An Evaluation of Computational Algorithms to Interface Between CFD and CSD Methodologies," WL-TR-96-3055, Flight Dynamics Directorate, Wright Laboratory, Wright-Patterson Air Force Base, Ohio, Nov. 1995.
23. Theodorsen, T., *General Theory of Aerodynamic Instability and the Mechanism of Flutter*, NACA TR 496, 1934.
24. Weisshaar, T. A. "Divergence of Forward Swept Composite Wings," *Journal of Aircraft*, Vol. 17, No. 6, June 1980, pp. 442–8.

Index

Printed in the United States
By Bookmasters